T0252217

The Metamorphosis of the World

The Metamorphosis of
the World

Ulrich Beck

polity

Copyright © Ulrich Beck 2016

The right of Ulrich Beck to be identified as Author of this Work has been
asserted in accordance with the UK Copyright, Designs and Patents Act 1988.

First published in 2016 by Polity Press
This paperback edition © Polity Press, 2017

Polity Press
65 Bridge Street
Cambridge CB2 1UR, UK

Polity Press
350 Main Street
Malden, MA 02148, USA

All rights reserved. Except for the quotation of short passages for the purpose
of criticism and review, no part of this publication may be reproduced, stored
in a retrieval system, or transmitted, in any form or by any means, electronic,
mechanical, photocopying, recording or otherwise, without the prior
permission of the publisher.

ISBN-13: 978-0-7456-9021-6
ISBN-13: 978-0-7456-9022-3 (pb)

A catalogue record for this book is available from the British Library.

Library of Congress Cataloging-in-Publication Data

Beck, Ulrich, 1944-2015, author.
  The metamorphosis of the world / Ulrich Beck.
    pages cm
  Includes bibliographical references and index.
  ISBN 978-0-7456-9021-6 (hardback) – ISBN 978-0-7456-9022-3 (pbk.)
1. Social change–Environmental aspects.   2. Climatic changes–Social aspects.
3. Risk–Sociological aspects.   4. Environmental sociology.   5. Social evolution.
I. Title.
  HM856.B43 2016
  303.4–dc23

                                                      2015032724

Typeset in 11 on 13 pt Sabon
by Toppan Best-set Premedia Limited

The publisher has used its best endeavours to ensure that the URLs for external
websites referred to in this book are correct and active at the time of going to
press. However, the publisher has no responsibility for the websites and can
make no guarantee that a site will remain live or that the content is or will
remain appropriate.

Every effort has been made to trace all copyright holders, but if any have been
inadvertently overlooked the publisher will be pleased to include any necessary
credits in any subsequent reprint or edition.

For further information on Polity, visit our website: politybooks.com

Contents

Contents

Foreword

The Story of an Unfinished Book

1 January 2015 was a splendid winter day: blue sky, sun all over, snow sparkling with light. It was scenery out of a picture book, filled with magic. In a joyful mood, Ulrich and I set out for a walk in the nearby park, Munich's famous *Englische Garten*. Some weeks before, at the beginning of December, Ulrich had sent a preliminary and unedited version of *Metamorphosis* to Polity Press, and just some two or three days previously, at the end of December, he had received the first reviews. While initially he had been somewhat irritated by some of the comments, now, in the course of our walking and talking, he came to see that they did indeed touch upon important issues. Immediately he started on a course of brainstorming, and I joined in. We spoke of adding new parts that would help to clarify and improve critical issues.

But then, in the midst of our brainstorming, the end.

A sudden heart attack.

Ulrich died.

A few days later, I tried to record the major points we had been talking about on that beautiful New Year's day. But, try as hard as I could, I could not accomplish the task.

Memory failed me. All I could remember were fragments, bits and pieces. The essence was gone.

In February 2015, the LSE paid a special tribute to Ulrich. At an event held in his honour, Anthony Giddens spoke of *Metamorphosis*, calling it an 'unfinished book'. In the following months I learnt the truth of his statement. This was when the task of transforming the preliminary manuscript into a book began and kept me going and going. It was but the last chapter in a long story that involved many people and was closely associated with Ulrich's ERC Advanced Grant 'Methodological Cosmopolitanism – In the Laboratory of Climate Change'.

From the very beginning Anders Blok (Copenhagen) and Sabine Selchow (London) had been engaged in discussing first drafts of the manuscript. Both Blok and Selchow, in their own way, have given much time, energy and expertise to this task. Thanks to their efforts, the manuscript gained in depth and theoretical foundation as well as in precision and empirical detail. Furthermore, numerous people – some also members of the ERC team, some colleagues from a variety of academic backgrounds, some based in Munich, some living in faraway regions and continents – have come up with fruitful suggestions and inspired new ideas. The following persons took part in this network of cosmopolitan cooperation: Martin Albrow (London); Christoph Lau (Munich); Daniel Levy (New York); Zhifei Mao (Hong Kong); Svetla Marinova (Sofia); Gabe Mythen (Liverpool); Shalini Randeria (Vienna); Maria S. Rerrich (Munich/ Blackstock, South Carolina); Natan Sznaider (Tel Aviv); John Thompson (Cambridge); David Tyfield (Lancaster/ Guangzhou, China); Ingrid Volkmer (Melbourne); and Johannes Willms (Munich). Once again Almut Kleine (Munich), trained by twenty years of working with Ulrich, bravely navigated through his handwritten notes and corrections and typed her way through many versions of the text. And Caroline Richmond at Polity did a wonderful job copyediting the text and ironing out any wrinkles that remained.

But, before that, there was the task of completing the unfinished book. It was a challenge indeed, and it needed the collaboration of three individuals.

Fortunately, as Ulrich and I had been close companions and colleagues for so many decades, the subject of metamorphosis had been part of our daily discussions – indeed, our daily lives. I had seen Ulrich struggling with it and eventually coming to terms with it. Furthermore, I could build on the experience of four books and numerous articles we had written together. Yet, when it came to producing a final version of *Metamorphosis* – a version ready for print – each chapter presented a series of open questions, from metaphors of mysterious meaning to arguments based on unknown sources. In such moments – and there were many of them – John Thompson, close colleague and most loyal friend, stepped in, investing enormous amounts of time and energy, of sociological knowledge and publishing experience. Whenever I longed for a break, for some time off from *Metamorphosis*, or even for a chance to finish my own book, John patiently brought me back in line, pressed me to go on, or went ahead himself. Time and again, he helped to make sense of and revise incomplete sentences, paragraphs that ended abruptly, and text (written in English) that sounded too German.

But, in the end, both John and I would have been at a loss if it had not been for Albert Gröber, scientific coordinator of the ERC team and noted expert on every detail of Ulrich's writings. During the difficult period directly following Ulrich's death, Albert did not only have a major role in steering the project through serious problems; at the same time he also actively contributed to finishing *Metamorphosis*. He ingeniously tracked down references, unearthed remote quotations, and compiled a list of relevant authors and publications.

In this way the unfinished manuscript gradually gained in shape and was eventually transformed into a book. I am deeply indebted to John and Albert, and my warmest thanks go to both of them.

I hope that, taken all together, we have done well, at least on most occasions. I hope the result allows us to see the vision Ulrich had in mind when he started on the journey to *Metamorphosis*.

Elisabeth Beck-Gernsheim
September 2015

Preface

The world is unhinged. As many people see it, this is true in both senses of the word: the world is out of joint and it has gone mad. We are wandering aimlessly and confused, arguing for this and against that. But a statement on which most people can agree, beyond all antagonisms and across all continents, is: 'I don't understand the world any more'.

The aim of this book is to try to understand and explain why we no longer understand the world. To this end, I introduce the distinction between change and metamorphosis or, more precisely, between change in society and metamorphosis of the world. Change in society, social change, routinizes a key concept in sociology. Everyone knows what it means. Change brings a characteristic future of modernity into focus, namely permanent transformation, while basic concepts and the certainties that support them remain constant. Metamorphosis, by contrast, destabilizes these certainties of modern society. It shifts the focus to 'being in the world' and 'seeing the world', to events and processes which are unintended, which generally go unnoticed, which prevail beyond the domains of politics and democracy as side effects of radical technical and economic modernization. They trigger a fundamental shock, a sea change which explodes the anthropological constants of our previous existence and understanding of the world. Metamorphosis in this sense means simply that what was unthinkable yesterday is real and possible today.

We have been repeatedly confronted with metamorphoses of this magnitude in recent decades, in a series of (in collo-quial terms) 'insane events', from the fall of the Berlin Wall, the September 11 terrorist attacks, catastrophic climate change all over the world, the Fukushima reactor disaster, and the financial and euro crises to the threats to freedom by totalitarian surveillance in the age of digital communica-tion brought to light by Edward Snowden. We are always confronted with the same pattern: what was ruled out before-hand as utterly inconceivable is taking place – as a global event, mostly observable in every living room in the world because it is transmitted by the mass media.

Part I

Introduction, Evidence, Theory

1

Why Metamorphosis of the World, Why Not Transformation?

This book represents an attempt to rescue myself, and perhaps others too, from a major embarrassment. Even though I have been teaching sociology and studying the transformation of modern societies for many years, I was at a loss for an answer to the simple but necessary question 'What is the meaning of the global events unfolding before our eyes on the television?', and I was forced to declare bankruptcy. There was nothing – neither a concept nor a theory – capable of expressing the turmoil of this world in conceptual terms, as required by the German philosopher Hegel.

This turmoil cannot be conceptualized in terms of the notions of 'change' available to social science – 'evolution', 'revolution' and 'transformation'. For we live in a world that is not just changing, it is metamorphosing. Change implies that some things change but other things remain the same – capitalism changes, but some aspects of capitalism remain as they have always been. Metamorphosis implies a much more radical transformation in which the old certainties of modern society are falling away and something quite new is emerging. To grasp this metamorphosis of the world it is necessary to explore the new beginnings, to focus on what is emerging from the old and seek to grasp future structures and norms in the turmoil of the present.

Take climate change: much of the debate about climate change has focused on whether or not it is really happening and, if it is, what we can do to stop or contain it. But this emphasis on solutions blinds us to the fact that climate change is an agent of metamorphosis. It has already altered our way of being in the world – the way we live in the world, think about the world, and seek to act upon the world through social action and politics. Rising sea levels are creating new landscapes of inequality – drawing new world maps whose key lines are not traditional boundaries between nation-states but elevations above sea level. It creates an entirely different way of conceptualizing the world and our chances of survival within it.

The theory of metamorphosis goes beyond theory of world risk society: it is not about the negative side effects of goods but about the positive side effects of bads. They produce normative horizons of common goods and propel us beyond the national frame towards a cosmopolitan outlook.

But the word 'metamorphosis' must still be handled gingerly and placed within quotation marks. It still bears all the hallmarks of a foreign body. Certainly, for the time being this word will probably have to be content with guest worker status, and it remains open whether it will ever become part of our common sense. At any rate, with this book I propose to adopt the migratory concept 'metamorphosis' into the social common sense of countries and languages. This is simply an attempt to offer a plausible answer to the urgent question 'What world are we actually living in?' My answer is: in the metamorphosis of the world. However, this is an answer that requires willingness on the part of the reader to risk the metamorphosis of their worldview.

And of course there is a second overwhelming term in the title: 'world', which is closely linked to the term 'humanity'. What is this about?

The talk of the failure of the world focuses attention on the concept 'world'. All institutions are failing; no one and nothing is decisive enough in confronting global climate risk.

And it is precisely this insistence on failure that is making the world the point of reference for a better world.

In this way, the concept 'world' has become familiar. It has become indispensable for describing the most mundane things. It has lost its aloof isolation, its Himalaya-like grandeur, and through the back door it has crept into and ensconced itself in our everyday, most private language. Nowadays, pineapples, no less than the nursing staff for the elderly, have a global background (and everyone knows this). Someone who asks where the pineapples come from receives the welcome information that they are 'flown-in pineapples'. Correspondingly, there are 'flown-in mothers', who want to (or have to) care and provide for other people's children here and their own children back home in accordance with the rules of 'long-distance love'. Even cursory reflection shows that the concepts 'world' and 'one's own life' are no longer strangers. They are now and henceforth bound together in 'cohabitation' – in 'cohabitation' because there is no official authenticating document (whether of science or the state) for this lifelong global union.

Having said all this, the question remains: Why *metamorphosis* of the world, why not 'social change' or 'transformation'?

Taking the Chinese case, transformation means what China has experienced since the Cultural Revolution and the Chinese economic reform: an evolutionary path from closed to open, from national to global, from poor to rich, from isolated to more involved. Metamorphosis of the world means more than, and something different from, an evolutionary path from closed to open; it means epochal change of worldviews, the refiguration of the national worldview. Yet, it is not a change of worldviews that is caused by war, violence or imperial aggression but one that is caused by the side effects of successful modernization, such as digitalization or the anticipation of climate catastrophe to humankind. The institutionalized national-international *Weltbild*, the world picture, the significance in how humans today apprehend the world, has withered. 'World picture' means that for every *cosmos* there is a corresponding *nomos*, combining normative and empirical certainties as to what the world, its past and its future, is all about. These 'fixed stars', fixed certainties, are not fixed any more. They are metamorphosed in a sense that can be understood as the 'Copernican Turn 2.0'.

Galileo discovered the fact that the sun is not circulating around the earth, but that the earth travels around the sun. Today we are in a different but somewhat similar situation. Climate risk teaches us that the nation is not the centre of the world. The world is not circulating around the nation, but the nations are circulating around the new fixed stars: 'world' and 'humanity'. The internet is an example of this. First, it creates the world as the unit of communication. And, second, it creates humanity by simply offering the potential of literally interconnecting everybody. It is in this space that national and other borders are renegotiated, disappear, and then built up anew – i.e., are 'metamorphosed'.

Consequently, 'methodological nationalism' is the lesson of the sun turning around the world or, to put it differently, the lesson of the turning of the world around the nation. 'Methodological cosmopolitanism', on the contrary, is the lesson of the earth turning around the sun or, better, the lesson of the nations turning around the 'world at risk'. From a national outlook the nation is the axis, the fixed star, around which the world turns. From a cosmopolitan outlook this nation-centric world picture appears historically false. The metamorphosis of the world means that the 'metaphysics' of the world is changing.*

---

* *On metamorphosis*: The origin of the word is Latin via Greek – *meta* (change) *morphe* (form) – and changing form is the key (first found in English in 1530 relating to magic or witchcraft). The best matched synonym is transfiguration, not reconfiguration. Thus the notion of 'metamorphosis' can be defined as a major change into something different and implies a complete transformation into a different type, a different reality, a different mode of being in the world, seeing the world and doing politics.

To understand why the world picture is 'historically false' we need to distinguish between the Copernican Turn in the natural scientific and in the social scientific 2.0 sense. The world picture which claimed that the sun is turning around the earth has always been false. It is only that this reality has been denied by those following and defending the religious dogma. The Copernican Turn 2.0 unfolds in reality – that is, in everyday activity – in the actual upheaval and downfall of the world order. This does not mean, however, that nations and nation-states dissolve and disappear but that nations are 'metamorphosed'. They need to find their place in the digital world at risk, in which borders have become liquid and flexible; they need to (re)invent themselves, turning around the new fixed stars of 'world' and 'humanity'.

In a similar manner to how the modern international world order, the sovereign state, industrialization, capital, classes, nations and democracy set in and unfolded after the collapse of the religious world order, global climate risk contains a sort of navigation system for the threatened world (see later). Climate risk denotes the path. But this does not mean that it will be a successful path. It is possible that humanity may choose a path at the end of which lies its self-destruction. This possibility exists not least because, when this path comes into plain view, it becomes clear that the 'eternal certainties' of the national worldview are short-sighted and wrong and lose their self-evidence as the beliefs of a whole epoch.

The history of metamorphosis is a history of ideological conflicts (wars of religion) – in the past regionally, today globally. We are experiencing a struggle between competing images of the world involving fierce, brutal conflicts, bloody conquests, dirty wars, terror and counter-terror – for example, Christians against barbaric heathens. Charlemagne built his Christian empire in the secure knowledge that it was permissible to kill for the holy faith, to wipe out the unbaptized and their culture. In an alliance with the pope, he imposed God's commandments with brute force. This

Christian-religious worldview was based on the unity of conquest and mission, on the alliance between the sword and the cross. Christian baptism was realized with violence in the act of subjugation. This religious worldview taught that peace is possible only as peace within the unity of Christendom.

In a historical variation on Galileo's discovery, the world no longer revolves around the minor princedoms, around the conflict between Catholics and Huguenots, between colonial masters and barbarians, between superhumans and subhumans. The race-centric view of the world is defunct (especially in Germany and in Europe as a response to the racial fanaticism of the Nazis) – the patriarchal world picture, too (though not in all parts of the world), and the world picture which proclaims equality but excludes women, slaves and 'barbarians'. Just take the founding fathers of the United States of America and its constitution, who did not even notice that African Americans were excluded from human rights – they took it as the most natural thing on earth.

And, again, what does 'withered' mean? Many, most likely even all of these world pictures still exist today simultaneously and alongside each other. 'Withered' means two things: first, the world pictures have lost their certainty, their dominance. Second, nobody can escape the global. This is because, as we will see in the chapters to follow, the global – i.e., the cosmopolitized reality – is not just 'out there' but constitutes everybody's strategic lived reality.

In order to grasp this, it is necessary to distinguish between *Glaubenssätze*, 'doctrines', and *Handlungsräume*, 'spaces of action', which are the existential parameters of social activity when it comes to world pictures. *Doctrines* can be particular or minority-oriented e.g., anti-cosmopolitan, anti-European, religiously fundamental, ethnic, racist; *spaces of action*, on the contrary, are inevitably constituted in a cosmopolitan way. The anti-Europeans actually sit in the European parliament (otherwise they don't matter at all). The religious anti-modernist fundamentalists celebrate the beheading of their Western hostages on digital channels and digital media platforms in order to shock the world with their inhuman terror regime. If tomorrow a group appears that propagates the political superiority of left-handed redheads, they will announce and practise their belief not just locally but globally.

Even immobile people are cosmopolitized. People who have never left their villages, let alone ever boarded a plane, are still closely and commonly linked with the world: in one way or other they are affected by global risks. And they are linked with the world not least because the mobile phone has come to be an integral part of the everyday across the globe. The metamorphosis in this, however, is not simply that everybody is (potentially) interlinked but that this entering into the 'world' means to enter something that follows a completely different logic. They end up in a world that is fundamentally different from what they think and expect – i.e., a world in which, as mobile phone users, they are metamorphosed into (data) resources and transparent and controllable consumers for global transnational corporations. This is a key feature of metamorphosis.

No matter if you want to save money by avoiding taxes or if you are infertile but long for a child, to reach your goal you need to understand and make use of the legal and economic differences that exist between various economic and legal realms in different national contexts. A developer who thinks strictly nationally – i.e., rigorously dismisses cheap foreign labour in favour of more expensive German building workers – will go bankrupt. To put it differently: those who take the national imperative as the imperative for their action – i.e., who stop at national borders – are the losers in the cosmopolitized world.

Of course, everybody is free to choose not to board an aeroplane or not to send emails. Yet, this decision means that they exclude themselves from the spaces of successful action. The world order arises from the historical necessity of acting beyond and across borders in order successfully to pursue fundamental goals in life. In other words, an imperative of cosmopolitized action arises globally: no matter what one thinks and believes – nationalistic, religious fundamentalist, feminist, patriarchal, (anti-)European, (anti-)cosmopolitan, or all of this together – if one acts nationally or locally one is left behind. Regardless of which past era people take flight to in thought – the Stone Age, the Biedermeier era, the time of Muhammad, the Italian Enlightenment or the nationalism of the nineteenth century – if their actions are to be successful, they must build bridges to the world, to the world of the 'others'. At the beginning of the twenty-first century, spaces of action are cosmopolitized, which means that the frame of action is no longer only national and integrated but global and disintegrated, containing the differences between national regulations in law, politics, citizenship, services, etc.

In the cosmopolitized world, even national elections are organized in a cosmopolitan way: the parties who want to win need to secure the votes of citizens abroad – e.g., Turks living in Germany, US citizens outside the United States. States that react to the 'cosmopolitan criminal' only nationally fundamentally miss the cosmopolitization of criminality. Only if one looks at and understands the cosmopolitized spaces of action of criminals and 'trans-legally' acting corporations is an adequate reaction and handling possible.

This is the end of cosmopolitan idealism and the beginning of the cosmopolitan realism of successful action. You have to open up for the world if you want to succeed!

For those for whom the nation, ethnicity or religion constitutes their metaphysical certainty, the world breaks down. Their despair makes them turn to national and religious fundamentalism. Consequently, hundreds of sociological studies, asking what people have in *mind*, tell us the story of a backlash to renationalized orientations. This might be true in relation to what people think, but what about their activities? Those studies focus only on orientations, thereby missing the essential point: whatever people think and believe in, they cannot escape the *Paradox of Metamorphosis* that *is* the cosmopolitized world: in order to defend their national and religious fundamentalism they need to act – in fact, *think* and *plan* – in a cosmopolitan way. Hence, they foster what they originally set out to fight against: the metamorphosis of the world.

If the poor do not act transnationally – i.e., if they don't become 'world mobile', in the sense of migrate – they risk getting poorer. The poor become poorer because they remain in the slums of Bangladesh, North Africa and the ghettos in the US. The rich become richer because they invest their money wherever they make more profit and can avoid paying taxes. This logic is true even for the social sciences: those who practise methodological nationalism will lose. Sociologists who do research only from within and about the national context block their careers and remain what they are: national sociologists.

If you want to be successful you need to discover yourself as an actor in cosmopolitized spaces of action. (This is a necessary but not a sufficient condition.) Let's take the example of the desire to have a child: you need to 'google' to find the right egg donor mother, surrogate mother or sperm donor. The same goes for household help, university degrees, job openings – everything needs to be 'googled' so that successful action is possible. It is the *cosmopolitan frame* that makes *local* action successful: just think of pineapples or the football club Bayern Munich!

Consequently, the distinction between doctrines and spaces of action is crucial: at the beginning of the twenty-first century the world is turning schizophrenic in a fundamental sense. Notwithstanding what people believe in, hope for or question, they have to act in a cosmopolitan manner if they want to be successful – in the economy, religion, the nation, the community or in their family, their jobs, their football club, their romantic life – last but not least, in their terrorism. Cosmopolitization includes the body as well. Those who eat only locally will starve. In fact, in times of climate change, those who just want to breathe local air will suffocate.

1   Conceptual Clarification: Cosmopolitized Spaces of Action

If you ask for the systematic characteristics of the notion 'cosmopolitized spaces of action', a number of constitutive aspects appear. In exploring these characteristics it is essential to keep in mind that the concept 'cosmopolitized spaces of action' is interlinked with the notion 'metamorphosis of the world'.

1 It is useful to distinguish between *action*, which combines reflection, status and perception held by actors, and *cosmopolitized spaces of action*, which exist even if they are not perceived and used by actors. To be clear, 'cosmopolitized' comes out of the theory of 'cosmopolitization' and is not to be confused with 'cosmopolitan', which refers to 'cosmopolitanism' as a norm. Beyond perceptions of actors (governments, business, religions, civil movements, individuals, etc.) there has to be an analysis of cosmopolitized spaces of action, which need to be understood as not institutionalized in a national frame. They are *not* integrated, *not* limited and *not* exclusive. They include transnational, transborder resources for action, such as the differences between national judicial regimes, radical inequalities and cultural differences.

This nexus of doings beyond borders and beyond taboos is not necessarily a value or emotional nexus but is often based on 'mutual ignorance' (surrogate mothers, kidney donors and kidney transplant recipients). In order to make use of them you do not need to have the corresponding passport, speak the corresponding language or have the corresponding identity. The differences make the difference! The differences between cultural traditions, the differences between rich and poor populations, the differences between law systems and the differences in geography constitute the new cosmopolitized structure of *opportunities*.

It is also necessary to distinguish between *actions* and *practices*. Practices are routinized, actions are reflexive, bridging and using transborder differences. They are the outcome of historical processes of learning by doing. They create *cosmopolitan milieus*, not only at the top and the middle of society but also at the bottom. Undocumented migrants become *Artisten der Grenze*, 'artists of borders'.

This does not mean that under certain conditions cosmopolitized spaces of action might not be turned into routinized 'fields of practices' (Bourdieu 1977, 1984) – i.e., that borders are redrawn and new systems of regulation are created and implemented. But the point is that cosmopolitized spaces of action are open opportunities of action which are subject not to the logic of reproduction but to the logic of metamorphosis of the social and political order.

2 In order to understand the nature of the cosmopolitized space of action, we need to understand the idea of *spaces of spaces*. Spaces of spaces open up unexpected opportunities, thereby making metamorphosing orders and cultural relativisms of law, values and state authority visible and usable. Obstacles (in the national frame) metamorphose into opportunities (in the cosmopolitan frame). Because foreign law allows what your country's law prohibits; because you are rich and can afford to buy organs while people in other parts of the world are so poor they have to sell them; because you can mobilize friends or fighters by internet communication, Facebook, etc. – for such reasons your political aims, your hopes and goals in life, may be fulfilled in the cosmopolitized spaces of action that are constituted in highly different ways. The experience of the relativity of values and prohibitions changes into a question: What is common practice in the US and Israel can surely not be a crime

over here, so why is it prohibited? Are our laws wiser than others? The pro and contra of argument and counter-argument renders all points of view suspect; each helps to undermine the other. Many people gain the impression that no one has a monopoly on the truth. That in turn raises the question: If all the opposing views seem to be well founded, how can there be an acceptable prohibition? The effect of these disagreements is to undermine the law's effective claim to legitimacy, so that people justify 'their' right to break the law by getting somewhere else what is prohibited here. What we see in the cosmopolitized spaces of action is the metamorphosis of value relativism into the legitimation of the prohibited.

In this sense, the idea of 'spaces of spaces' differs fundamentally from Bourdieu's 'fields of fields', because the latter exist in the unity of the nation-state. Spaces of spaces include exclusive national fields of practices. In contrast to my notion of 'cosmopolitized space of action', Bourdieu's influential notion of 'fields of practices' makes sense of how broader structures of social and cultural domination are lived, reproduced and transformed in everyday life and practice or practices (methodological nationalism).

3 In order to understand 'cosmopolitized action', it is useful to bring in the concept of 'creative action' (Joas 1996). 'Creative action' is about the ability not to accept existing borders of thinking and acting. Even more than that, one needs to be ready and able to translate existing borders into opportunities in order to achieve one's goals. The creativity of cosmopolitized action means that the rationality of action metamorphoses. The notion of 'rationality' is metamorphosed because of the 'simple' fact that the internalization of the world has become the condition for successful action.

4 A key characteristic of 'cosmopolitized spaces of action' is that they do not equal particular ways of

thinking, doctrines, religious beliefs and ideologies. Rather, they are used strategically; in fact, they have to be used strategically if one wants to be successful – that is, if one wants to achieve one's goals. National elections are a good example. It might not be a successful move to follow a normative cosmopolitan *doctrine*, but there is no way around strategically *acting* in and through the 'cosmopolitized spaces of action'. There are different ways of doing this, the most prominent of which is strategically to instrumentalize the cosmopolitized resources behind a national façade.

5 For the first time in history there is a space of action that is open to everybody. In fact, from now on it is an *active* decision *not* to use the cosmopolitized spaces of action (or spaces of cosmopolitan resources for action). They are not exclusive in the sense that only powerful economic, political and military actors can make use of them. Individual actors can also use cosmopolitized resources – depending on their social position and economic means. It also implies a chance for 'upward mobility'. The cosmopolitized resources can be used by people living 'at the bottom' through enforced migration, which enables them to use the ladder to rise to a better life, even if the outcome is a mixture of disappointment and despair. This means that the situation is fundamentally different from one in which the cosmopolitized spaces of action do not exist, as has been the case in the history of mankind up to the last part of the twentieth century.

Today we all are global players, more or less! Maybe not voluntarily, maybe not deliberately, but because the cosmopolitan spaces of action offer superior chances of success compared to nationally, religiously and ethnically limited action in the cosmopolitized world. We know what Erd*anziehungskraft* is – the gravitational pull of the planet earth. This book unveils, reveals, and thinks through the new historical law of Welt*anziehungskraft* – the gravitational pull of the world.

2 Conceptual Clarification: the Notion of Metamorphosis

The metamorphosis of the world is apparent not least in the way in which the dominant cultural pessimism is becoming metamorphosed. Nowadays, many people see the preachers of catastrophe as the last remaining realists. They believe that the catastrophists' pessimism offers the best arguments when it comes to making a robust assessment of the situation:

> It's only a matter of time until this planet is so profoundly convulsed that we will fly away from it like annoying insects. The gentle twitches that we are already experiencing are merely the seismic harbingers of a global collapse which – if you believe the credible preachers of catastrophe – has become irrevocable. Under these circumstances it is hardly surprising that everywhere small competing groups are forming that offer their homeopathic healing arts as a way of saving the world: everything a little smaller, if you please, more credible, more manageable, fairer, simpler, cleverer, more human. Everyone of goodwill agrees wholeheartedly with them – only, please, not right now, not here…in Germany, in Europe, but first of all over there, where I am not right now. The rescue of the world is always supposed to start somewhere else, where the individual is not. (Krüger 2009)

We all know that the caterpillar will be metamorphosed into a butterfly. But does the caterpillar know that? That is the question we must put to the preachers of catastrophe. They are like caterpillars, cocooned in the worldview of their caterpillar existence, oblivious to their impending metamorphosis. They are incapable of distinguishing between decay and becoming something different. They see the destruction of the world and their values, whereas it is not the world that is perishing, but their image of the world.

The world is not perishing as the preachers of catastrophe believe, and the rescue of the world, as invoked by the optimistic advocates of progress, is not imminent either. Rather, the world is undergoing a surprising, but understandable, metamorphosis through the transformation of the reference horizon and the coordinates of action, which are tacitly assumed to be constant and unchangeable by the aforementioned positions.

The negation of pessimism does not imply optimism. This book is not about being an optimist or a pessimist, but about unpacking the dystopian and pessimistic constellation by way of identifying its sociological, political and cultural roots and conditions. We are totally confused because what was unthinkable yesterday is possible and real today on account of the metamorphosis of the world: yet, in order to be able to grasp this metamorphosis in a meaningful way, it is necessary not just to explore the dissolution of socio-political reality but to focus on new beginnings, on what is emerging and on future structures and norms.

As I said before, the Copernican Turn 2.0 means that the imperative to conceive of and affirm the nation as the fixed star around which the world rotates is being supplanted by the imperative to think of the 'world' and 'humanity' as fixed stars around which the nations rotate. How, in what forms and steps, is this metamorphosis of our worldview taking place? Not as an ideological-cosmopolitan programme from the top down, as the philosophical textbooks would have it. Rather, the agent of metamorphosis of the world is the endless story of failure. To put it bluntly, global poverty is on the rise, the poisoning of the planet is on the rise, so too is global illiteracy, while global economic growth leaves much to be desired, the world's population is growing ominously, global famine relief is inadequate, and the global market – especially the global market – is driving us all to ruin. This persistent public lament is what initiates and hammers home the change in worldviews. Crucial in this regard are not simply statistics as such, but that they are communicated publicly as a scandal, as an outrageous political and moral failure. In this way, the notions of 'world' and 'humanity' are rendered plausible as ultimate points of reference, as the new fixed stars, and are produced and reproduced as a structure of rationality. Through the television images of the everyday consternation over the failure of institutionalized action, the old social and political order is being metamorphosed and first steps are being made towards the production and reproduction of a new order, now literally a 'world' order. The paradox is that the complaints and accusations about the failure of the world are awakening the consciousness of the world.

This is the theme of an empirical sociology of the metaphysics of the worldview metamorphosis, something that I can only hint at here.

As we know, theoretical concepts frequently invite misunderstandings, which then provide the material for controversies that fill entire libraries. This will doubtless be the case with the concept of the 'metamorphosis of the world' presented here. To forestall such possible misunderstandings we shall now attempt to define the concept more precisely.

Normative political vs. descriptive

When sociologists speak of 'change' (or 'social change'), this is often understood as *political* change – in other words, as a programmatic change in society beneath the banner of socialism, neoliberalism, fascism, feminism, colonialization, decolonialization, Westernization, etc. Such a consciously intended, programmatic change in society with specific goals in mind is precisely what is *not* meant by the concept of the metamorphosis of the world. The metamorphosis of the world is something that happens; it is not a programme. 'Metamorphosis of the world' is a descriptive expression, not a normative one.

Everything or the new

If in what follows I am concerned to introduce this concept of the metamorphosis of the world, this does not mean that

I think of *everything* that occurs in society today – in the economy and in politics, the world of work, the education system and the family, etc. – as a metamorphosis. That is certainly not my intention. Such a global assertion would be exaggerated and also false. But, by the same token, it would be equally mistaken to leave metamorphosis out of account from the very outset – as is customary in traditional theorizations – and to refuse to see it even as a possibility.

From my vantage point, by no means everything can be said to be a metamorphosis of the world. On the contrary, we are looking here at the simultaneous presence, the intertwining, of the world, social change, and the reproduction of the social and political order with all its countervailing movements. I am concerned not with the present in its totality but with what is new in our present reality.

This is the crucial difference between my approach and the current theories and research routines of the social sciences, which are focused exclusively on social change within the framework of the reproduction of the social and political order. Their very approach precludes the possibility of the metamorphosis of the world. In contrast, my starting point is that it is only in the context of the metamorphosis of the world that we can explore the relations between metamorphosis, change, reproduction and its countervailing movements. The relative weighting of each of these factors is something that must be investigated empirically.

In short, by introducing the concept of the metamorphosis of the world, my intention is not to replace the existing typology of historical change in society and politics with a completely different one. My goal is to *complement* that typology with a new one that has hitherto passed unnoticed.

No determinism – whether optimistic or pessimistic

It would be no less misguided to equate the metamorphosis of the world with a change for the better. Metamorphosis of the world says nothing about whether a given transformation is for the better or the worse. As a concept, it expresses neither optimism nor pessimism about the course of history. It does not describe the decline of the West, nor does it suggest that all will be for the best. It leaves everything open and points us towards the significance of political decisions. It highlights the potential of the world risk society to lead to catastrophe, but also to the scope for an 'emancipatory catastrophism'.

Uniform vs. diverse metamorphosis of the world

By asserting that the metamorphosis of the world is the characteristic feature of the present age, I do not wish to imply that it will assume the same form in every region of the world. To take the example of climate change once again, it is well known that, while the melting of the glaciers can pose an existential threat to polar bears, for mankind the same process may create new opportunities for agriculture and oil exploration. Climate change may have different or even opposing consequences for different groups in the same region, and the same may be said with even greater force of different regions. Climate change may well lead to drought in one region and new vineyards in another. For this reason it is essential to focus on the social geography of the metamorphosis of the world. This gives rise to a complex multi-level model of metamorphosis that takes account of the interaction of local, regional, national and global conditions and develops specific structures as the consequence of social inequalities and social power relations.

In sum, metamorphosis is not social change, not transformation, not evolution, not revolution and not crisis. It is a mode of changing the nature of human existence. It signifies the age of side effects. It challenges our way of being in the world, thinking about the world, and imagining and doing politics. And it calls for a scientific revolution (as Thomas Kuhn [1962] understands it) – from 'methodological nationalism' to 'methodological cosmopolitanism'.

Metamorphosis of the world and the world risk society

The concept of the metamorphosis of the world that I am introducing here does not imply that we can conceive of only one specific form of metamorphosis. On the contrary, there can and will be various theories of the metamorphosis of the world, just as there are various theories of change, revolution and evolution.

In this book I wish to develop a specific theory of the metamorphosis of the world, namely one that arises from its connection with the theories of world risk society, cosmopolitanization and individualization – in other words, reflexive modernization and second modernity.

Diagnosis and description

But how can we operationalize and empirically test the validity of this connection between the concept of the metamorphosis of the world and the theory of the world risk society? Metamorphosis of the world is not assumed to be 'normal' in the way that 'change' or, in a different way, 'revolution' and 'evolution' are. It isn't normal in the statistical sense either. It is unknown territory. For that reason I develop in what follows a set of interconnected middle-range descriptive concepts of the metamorphosis of the world – such as 'cosmopolitized spaces of action', 'risk-class', 'conditions of definitional power', 'emancipatory catastrophism', 'cosmopolitan communities of risk', etc. In this sense, this book is a thought experiment to be scrutinized empirically in the ERC research project 'Methodological Cosmopolitanism – in the Laboratory of Climate Change'.

2

Being God

The metamorphosis of the world, I argue, includes the metamorphosis of the world picture, which has two dimensions: the metamorphosis of framing and the metamorphosis of practice and acting. This idea will be developed further in the present chapter. The worldview always also contains an image of humanity. Taking the example of reproductive medicine, I propose to trace, on the one hand, the metamorphosis of human life and, on the other, the metamorphosis of the image of humanity, the image of motherhood, fatherhood and parenthood that was valid for millennia. This means that a new cosmopolitan framework and space for action is emerging along with the new options presented by medical technology, in particular also where the old image of humanity still dominates people's thinking. To put it in a nutshell: what used to be an intimate and almost 'sacred' act has metamorphosed into a global cosmopolitized field of activities.

1 Why not Social Change, Why Metamorphosis of Parenthood?

Throughout human history up to the present day, two things were regarded as unshakeable. First, it was impossible to control human reproduction (except for very unreliable contraception practices and the possibility of abortion). Second, care and responsibility for children was a moral law (even if one that was frequently violated).

Whether war or peace, master or servant, early or late modernity, centre or periphery, an indissoluble, predetermined relationship qua natural law runs through all phases, situations and groupings of human history – namely, the biological unity of mother and child that marks the beginning of human life.

This unity can assume many manifestations and can even be adapted to the most diverse ideologies and worldviews. In eighteenth- and nineteenth-century Europe, the mother was transfigured into a mythical figure and placed on the altar of motherly love in philosophy, religion and education. In the twentieth century, motherhood was instrumentalized in Nazi Germany for the purposes of world conquest and was rewarded with the *Mutterkreuz*. A few decades later, in the course of the expansion of higher education, rising female employment and powerful women's movements, motherhood became a major stake in cultural struggles: on the one side, the 'uncaring mother' who neglects her children, on the other, the 'urban housewife', the over-solicitous homemaker.

Today we can find a new plurality of mothering arrangements – employed mothers, single mothers, stay-at-home mothers. But the background assumption even in feminist studies is often that mothers and their children live in one place. In reality transnational motherhood is emerging: mothers migrate to faraway countries, leaving their children back home, in order to earn money and give them better options in life (Hondagneu-Sotelo and Avila 1997).

Some of these recent trends have been seen and described as dramatic. Nevertheless, all of them fall under the category of social change. While they represent major shifts in gender relations, the division of work between the sexes, and the position of women, they do not touch, interfere or meddle with the origins of human life. The metamorphosis of the

world in relation to motherhood and fatherhood, by contrast, begins with the malleability of conception by medical
technology. The genesis of human life is exposed to human
intervention and creative will, but as a result also becomes
the playground of the most diverse actors and interests scattered across the world (Beck-Gernsheim 2015).

What is taking place here cannot be understood as a
'crisis' of prenatal hominization or as a 'failure' of science
which must be overcome in order to return to the natural
process of procreation. Here, in the cooperation of medicine,
genetics and biology and the successes resulting from this
cooperation, thresholds of mutability and of interest intervention are being crossed irrevocably, with in vitro fertilization playing a key role. This term refers to fertilization in a
test tube – IVF for short. It was performed for the first time
in 1978 in Britain, and it instantly became a medical sensa-
tion. For the first time in human history, a child was born
who had been conceived outside the womb.

2   Being God without Wanting to be God

What does metamorphosis mean here? A key is provided by
the side-effects argument. The original aim was to treat the
fertility problems of women – to be more precise, of wives
(because, initially, nobody gave a thought to the desire of
single women for children). In order to be able to perform
this repair task informed by the conventional picture of the
family, a more precise understanding of the functional processes in the area of fertility and infertility is necessary. This
increasingly more precise understanding gives rise, in turn,
as a side effect, to the possibility of increasingly more extensive interventions in the development of human life.

In other words, the pioneers of reproductive medicine
were not trying to change our image of humanity. They were
not driven by an ideology or a political programme, and they
were not trying to bring about a revolution. On the contrary,
their objective, as we must now observe in retrospect, was a
very conventional one. It was to help desperate couples to

have the child they so fervently desired by using medical technology to bypass a blocked fallopian tube. What was, biologically speaking, a major pioneering feat served initially in the social context to reproduce the traditional picture of the family. What could be more natural than to use a medical intervention to fulfil a deeply cherished 'natural' desire of married couples? The medical pioneers were far from wanting to play God, to act as lords of creation or to create the 'new man'. They simply wanted to enable desperate couples to fulfil their hope for a child.

But, however conventional this starting point, the discrepancy between thought and action is already apparent. While the pioneers' aims were still fixed in the framework of the old worldview and a traditional concept of the family, at the practical level the gates were thrown wide open to the ever more extensive manufacturability of human life. This is the first step towards the metamorphosis of the image of human beings and the world – or, to be more precise, of the framework for action with regard to conception, pregnancy and parenthood.

The second step is implicit in this technical horizon. The unity of conception, pregnancy and birth formerly stipulated by nature as a matter of fate in the person of the mother breaks apart, and these sub-processes become uncoupled in space and time and at the social level. This gives rise to new options, forms and relationships in the emergence of human life for which existing language still lacks appropriate words and concepts. The reason is obvious: all languages in the world are rooted in the old horizon of the pre-given unity of parenthood. The inflationary use of quotation marks testifies to this helpless attempt to capture in language what never existed before, what was previously unimaginable.

The act of procreation no longer occurs face-to-face or body-to-body in a personal, physical encounter between man and woman. It no longer calls for the presence of two people at the same time in the same place, but can be displaced into a laboratory somewhere in the world, in any random, rented womb at any arbitrary time. Indeed, more importantly, the biological 'father' and the biological 'mother' need not even live or have lived at the same time, because now even dead people can conceive and give birth to children.

This gives rise (independent of the intentions and self-understanding of reproductive physicians) to historically new, previously unknown options, and thus also social categories, of parenthood: 'social mothers' who 'order' and 'buy' a child; 'sperm donors' and 'egg donors' who sell the biological 'materials' for 'making' a child; 'surrogate mothers' who carry a child; 'fatherless mothers'; 'motherless fathers'; post-menopausal 'pregnant' women; 'gay fathers'; 'lesbian mothers'; mothers and fathers whose partners are (long since) dead; grandparents who have a grandchild conceived after the death of their son or daughter; and more of the like.

These linguistic formulas are all inadequate, misleading, controversial, provocative – even, for some, offensive. They reflect the breach of taboos prompted by the medical-technological manufacturability of human life. The method of rendering the new reality of parent–child relationships tangible and comprehensible by falling back on familiar concepts truncates and normalizes the process of metamorphosis that has been set in motion.

Another wave of side effects (metamorphosis) arises because the technical innovations mentioned coincide with a rapid transformation of lifestyles and family forms in Western societies. The result is that the set of potential clients of reproductive medicine has expanded enormously within the space of a few years. With the social normalization and legal recognition of forms of life and lifestyles that were previously taboo, targets of discrimination or even criminalized, new groups are now also declaring their desire to have children: unmarried couples, singles, gays and lesbians, post-menopausal women, and so forth (Beck-Gernsheim 2015: 98–9). Now that the basic right to equality also applies to these groups and that, simultaneously, the range of options provided by medical technology for fulfilling the desire to have children is expanding rapidly, there is no longer any reason in principle to withhold the corresponding options from these lifestyle groups – with the result that the dams are bursting.

In reality, however, there are two major, secondary barriers that severely restrict the use of what is technically feasible. First, the corresponding forms of treatment are technically elaborate and therefore very expensive. Second, the medical-technological options and the possibilities for using them are perceived and evaluated in very different, often even conflicting, ways in the context of different religious and cultural worldviews and conceptions of the human being (Beck-Gernsheim 2014; Inhorn 2003; Waldman 2006). A cross-country comparison, therefore, reveals in practice very different legal regulations and religious prescriptions, ranging from laissez-faire (USA, Israel) to sweeping restrictions (Germany).

3   Prenatal Cosmopolitization

The cosmopolitized world offers special possibilities for dealing with the problem of high costs. Because medical technology has disembedded, objectivized and specialized conception, pregnancy and birth, these can now be distributed and reorganized according to the principles of economic rationality and the rules of the global market. They become a field of activity of 'outsourcing capitalism' governed by the principles of cost minimization and profit maximization. They are being distributed across continents in accordance with the rules of global inequality and the global division of labour. Hiring a surrogate for the nine months of pregnancy is expensive in the wealthy regions and much cheaper in countries with a broad pool of poor women. In this way, the groundwork for a new global economic sector is also being laid. What is somewhat optimistically called 'fertility tourism' specializing in the 'commodity child' is starting, and thereby ultimately it is giving rise to the social figure of the prenatal cosmopolitan patchwork family.

The essence of capitalism resides in its dynamics, and specifically in its ability to overcome the existing obstacles to the metamorphosis of 'natural' motherhood into the industrial production of prenatal motherhood, thereby opening it up to global market exchange. This kind of 'prenatal cosmopolitization' begins with 'prenatal accumulation', the expropriation by fertility doctors and fertility clinics of the biological resources of conception from 'natural fathers' ('sperm donors') and 'natural mothers' ('surrogate mothers'). The sacredness of motherhood and the national restrictions on the global market exchange of those biological resources are being overthrown, because the global inequality between the rich and the poor minimizes costs and maximizes profits. Therefore 'prenatal capitalism' shifts the centre of gravity of social life – motherhood – from a traditional, biological, sacred unity into an 'invisible cosmopolitization', creating and integrating territorial and social forms of biological fatherhood and motherhood 'from afar' into the children's fates. As a result, prenatal childhood becomes the focus of global, legal, political, ethical and religious concern, debate and conflict.

What is taking place in the laboratories of reproductive medicine and in the clinics of the prenatal industry does not constitute a 'revolution', because it has nothing to do with political upheaval or regime change among elites. It can't be grasped in terms of the concept of 'evolution' either, because it doesn't follow an antecedent law of development or basic principle (biological selection, functional differentiation, etc.). The paradox of the metamorphosis of prenatal hominization can be put as follows: unintentionally, without a purpose, unawares, beyond politics and democracy, the anthropological foundations of the beginning of life are being reconfigured through the back door of the side effects of the success of reproductive medicine.

4   A New World and World Picture of Human Life is Emerging in the Shadow of Speechlessness

The point or even the paradox of metamorphosis is that, unseen and unintentionally, beneath the surface of our imagined eternal concepts of being human, a new world and world picture is emerging with the normative power of the factual, perhaps even eventually a new world order for which we do not have any concepts, for which we literally lack a language. An uprising rebels against this, now here, now there, and immediately loses its way again in self-reflexive speechlessness.

Metamorphosis, understood in this way as a global revolution of side effects in the shadow of speechlessness, triggers a chain reaction of the failure of institutions in the full splendour of their functionality (chapter 7). Politics (insofar as it even makes a claim to regulate) fails if only because, by its very concept, it can operate simply within national boundaries and antagonisms – but the global side-effects revolution in reproductive medicine escapes the nation-state's attempts at regulation. Law, along with the different conceptions of law, fails for the same reason. Finally, our understanding of morality and ethics also fails. On the one hand, the questions and alternatives posed by the prenatal malleability of the human condition receive very different, and at the extreme diametrically opposed, valuations in different traditional contexts and cultural spheres; on the other hand, relevant studies show that universal values such as protecting human dignity justify both the prohibition of and the injunction to use the prenatal technologies and alternative mode of configuring parenthood. *Whose* human dignity is supposed to be protected when the mothers and fathers who are often scattered across the globe are at the same time (biologically) included and (socially) excluded from the new 'family forms'?

This in turn reflects the worlds of difference between change and metamorphosis. Change occurs within the existing order and the anthropological certainties that support it, which are embedded and predetermined historically and institutionally in the forms of nation-state politics and law and in the notion of universal values (protecting human dignity). Metamorphosis destroys these, while at the same time placing the existing institutions under enormous pressure to act through new, previously unimaginable practical alternatives. This pressure, as indicated, cannot be mastered with the usual concepts and instruments. Accordingly, the result is a 'reformation' of the nation-state order of modernity. By 'reformation' I mean (following and differentiated from the Reformation set in motion by Martin Luther against the Catholic Church with his 'Here I stand; I cannot do otherwise') a meta-politics, a politics of politics, a politics that reshapes the nation-state understanding and corresponding norms and institutions – not in all possible directions and counter-directions, but with the aim of the cosmopolitan renewal and extension of the transformative potentials of national politics (chapter 9). This elicits embittered resistance at all levels and in all contexts by counter-reformers who defend the old certainties and their institutionalized order against the onslaught of a world that has become 'unhinged'.

The litmus test for this is conflict. Among the best-known cases of conflict are disputes between surrogate mothers and the contracting parents that break out when the surrogate mother wants to keep the child after birth, contrary to the contractual agreements, and the contracting parents bring a lawsuit for the surrender of the child. Who has a 'right' to the child in this case? To whom does the child belong? Who should be considered its mother or father? Such cases have kept the courts busy, sometimes for years. When the 'mothers' assert conflicting claims in court to 'their' child, to 'true motherhood', judges find themselves in the Brechtian chalk circle dilemma. However, unlike Brecht's judge, they cannot take the wisdom of life experience as a basis for their decisions but must adhere to the articles of the national law. The only question is: which law, which articles?

This question becomes particularly contentious in countries in which common law, which is based on precedents, applies. But where are such precedents to be found in an era where what has never existed before is suddenly a reality? Undoubtedly, nowadays reproductive medicine can help many men and women to fulfil their long-cherished desire for a child. But at the same time it also gives rise to human tragedies in which the interests and desires of new kinds of groups collide – 'contracting mother' against 'biological father', 'social father' against 'surrogate mother', 'social mother' against 'biological father', child against 'biological father', and so forth.

The reproduction industry operates globally; politics and law respond to the challenges nationally. But national legislation is being increasingly undermined at the global level by men and women diverting to countries with less stringent regulations. The result is a maze of regulations that overwhelms the staff of the relevant authorities. Thus, registrars in Germany or the staff of German embassies increasingly have to deal with German men, women and couples who, for example, engaged the services of a surrogate mother in India but, when they wanted to take the child back to Germany, became caught up in the contradictions between legal systems. Under Indian law, the child has German parents and is therefore not entitled to an Indian passport; but it cannot get a German passport either, because surrogacy is illegal in Germany and hence is not recognized by the German authorities. Therefore, the German federal association of registrars called for a reform of the legal situation some years ago. 'The Federal Association of German Registrars', in the words of its resolution, 'considers a reform of family law necessary due to the increase in surrogacy, which is prohibited in Germany.'

It is not only lawyers and civil servants, but all of us who are lagging behind, in our language and thought, the metamorphosis of the world, which is becoming a reality with the sudden possibility of manipulating the beginning of human life. We are all prisoners of a language that conserves the old certainties of motherhood and is blind, and makes us blind, to the new diversity of options and forms of parenthood. The womb is no longer the mother's womb – which mother? The fatherland no longer exists; rather, it is lands of the fathers. And whereas it used to hold that 'pater semper incertus', in the present age of genetic technology the legal formula is 'pater certus'. But, at the same time, the principle 'mater certa' no longer holds; instead, it is 'mater incerta' – the child has many mothers.

Concurrently, the slick term 'sperm donor' (by the way, a euphemism that disguises the commercial act of selling one's sperm) reduces the male to the raw material supplier of the reproduction industry and suggests a biological relationship beyond responsibility and ethics. But the untenability of this euphemism ultimately becomes apparent when the children of 'sperm donors' begin to ask about their origin and go in search of the great unknown, their 'biological father'.

5   Outlook: the Categorical Imperative of Parental Responsibility is Breaking Down

The example of reproductive medicine shows how nowadays people – even if they happen to live in a remote Turkish provincial city or in a Swabian village and have never left their birthplace – operate within a cosmopolitized field of action and are better able to realize their basic life goals and desires if they overcome the cultural and financial constraints of their national environment. Those who want a child at any cost must overcome local and national boundaries and use the possibilities afforded by the global space. They must explore and compare offers extending from Ukraine to India, seek out loopholes, and even be ready to take 'trans-legal' detours, and, if necessary, also to select options that are prohibited by the laws of their country or the rules of their religion.

As has been shown, a new image of humanity and the world is taking shape as a product and a side effect of the rapid developments in medical technology. This is occurring almost imperceptibly, in many small successive steps, but neither in the sense of a purposeful evolution nor in that of an ideologically defined and systematically planned revolution.

Metamorphosis of the world means, therefore, that the image of humanity which seemed to be fixed for all time is disintegrating and a new one is emerging of which we can at present discern only the initial imprecise contours. The controversies over reproductive medicine are ultimately always also tacitly about defending an old and imposing a new image of humanity.

In this debate, the protagonists argue that the outcome is what matters – the birth of the child justifies the means. On the other hand, critical voices point out that the fundamental questions thrown up by the manufacturability of human life are being answered through the mute power of the capitalist world market and are thus being suffocated, as it were, before they can even be posed and publicly debated.

While the old conception of what it means to be human was based on the categorical imperative of parental responsibility, this principle is being eroded by the technical differentiation, multiplication and anonymization of parenthood. If a child is born severely disabled, if the longed-for dream child 'accidentally' turns out to be quadruplets or quintuplets, if the contracting parents divorce or die – then who is responsible for the welfare of the child? Who decides what is lawful in this case, and on the basis of what law? Here, at the heart of the industrialized and globalized emergence of human life, a diffuse, highly controversial legal field is emerging, a no man's land of responsibility or irresponsibility.

Today we can find precursors of the provisionality of parental responsibility in the procedures for dealing with (in the revealing terminology) 'surplus embryos'. Should they be frozen? Be donated to other couples? Be made available for research purposes? Be sold for profit – according to the principle implant the 'good' in one's own womb, sift out the 'bad' and give them away?

Here, trapped in the technological jargon, we already hear talk of 'quality control' of embryos, 'stockpiling' of embryos, and so forth, which fails to mention that this involves pre-natal selection and, if necessary, the destruction of future human life.

As a result, a *paradox of the new image of humanity* is becoming apparent, namely, that, precisely where the desire for children is so pressing and all-consuming, an indifference, indeed an organized irresponsibility, insinuates itself into the technical procedures and is practised, which becomes perfectly 'natural' and replaces the unconditionality of parental responsibility.

There is no sign of an 'emancipatory disaster' (chapter 7).

3

How Climate Change Might Save the World

Today, most discussions on climate change are blocked. They are caught by catastrophism circulating on the horizon of the problem: What is climate change *bad* for? From the point of view of metamorphosis, because climate change is a threat to humankind, we can and should turn the question upside down and ask: What is climate change *good* for (if we survive)? The surprising momentum of metamorphosis is that, if you firmly believe that climate change is a fundamental threat to all of humankind and nature, it might bring a cosmopolitan turn into our contemporary life and the world might be changed for the better. This is what I call *emancipatory catastrophism* (chapter 7; see also Beck 2015).

To avoid misunderstandings, I do not argue that we need a big-bang catastrophe in order to become reborn optimists, nor do I want to draw and advocate the counter-picture of a hyper-optimism, expecting a technological salvation from all evils of the present world by digital innovations (as some do). The cosmopolitan metamorphosis of climate change (or global risk in general) is about the co-production of risk perceptions and normative horizons: the apocalypse knows no constraints. Living in suicidal modernity (capitalism), the black box of fundamental political questions is reopening: Who speaks for 'the cosmos'? Who represents 'humanity'?

Is it the state? The city? Civil society actors? Experts? 'Gaia' (Latour 2011)? And who speaks for their own kind?

The global risk of climate change is a kind of compulsive, collective memory – in the sense that past decisions and mistakes are contained in what we find ourselves exposed to, and that even the highest degree of institutional reification is nothing but a reification that can be revoked, a borrowed mode of action which can, and must, be changed if it leads to self-jeopardization. Climate change is the embodiment of the mistakes of a whole epoch of ongoing industrialization, and climate risks pursue their acknowledgement and correction with all the violence of the possibility of annihilation. They are a kind of collective return of the repressed, wherein the self-assurance of industrial capitalism, organized in the form of nation-state politics, is confronted with its own errors in the form of an objectified threat to its own existence.

1   What Does Climate Change Do to Us?

Addressing climate change at the level of world (and local) politics, we may distinguish two basic framings of the issues involved. The first framing asks a normative and political question: 'What can we do against climate change?' This is the mainstream question posed by scientists, politicians and environmental activists looking for solutions to the problem, even as this proves disillusioning. By contrast, the second framing (informed by metamorphosis) poses the sociological and analytical question 'What does climate change do to us, and how does it alter the order of society and politics?' Posing this question allows us to think beyond apocalypses or the salvation of the world and focus on its metamorphosis. In this way it allows us to step back and rethink the fundamental concepts into which current discourses of climate politics are trapped and to explore the ongoing metamorphosis taking place under the radar.

Under the stress of finding workable solutions, the first question tends to dominate the second. This is one important reason why, at the present moment, our collective powers of social and political imagination seem blocked. The situation of blockage is compounded, however, by two additional factors. First, the sheer *success* of the predictive powers of the climate sciences now introduces a paradoxical situation, where public and media discussions of climate change happen under the guillotine of the 'tipping point' (cf. Russill and Nyssa 2009). Never before in human history has political life been saturated by this much knowledge about a pending global emergency. Rather than contributing to considerate public responses, however, the rhetoric of tipping points accelerates the issue and stands in the way of a socio-political rethinking.

Second, just at the time when the spectre of climate change stages the need for a large-scale POLITICS of the planet itself, global publics find themselves confronted with the sheer impotence of actually existing national-international politics. As witnessed in the drama enacted during the COP15 Summit in Copenhagen in 2009, the disconnect is indeed huge – and, following a massive build-up of societal expectations, the resulting political disappointments are equally deep. In place of the re-emergence of politics, an *apocalyptic imaginary* now dominates the public sphere, serving as an 'affective prophylaxis' meant pre-emptively to prevent overly strong traumatic shocks from the 'pre-mediated' catastrophe (Grusin 2010; Swyngedouw 2010). The climate pessimists promulgating this apocalyptic imaginary behave much like the famous angel of history in Walter Benjamin's parable about Paul Klee's picture *Angel of History*: the storm of climate change irresistibly propels them into a political future to which their backs are turned and which they remain unable to see or grasp.

In this book, I hypothesize that the main source of climate pessimism lies in a generalized incapacity, and/or unwillingness, to rethink fundamental questions of social and political order in the age of global risks. To counter such incapacities, the cosmopolitan theorizing and research that I promulgate hinges on the recognition that climate change alters society in fundamental ways, entailing new forms of power, inequality and insecurity, as well as new forms of cooperation, certainties and solidarity across borders. Three facts illustrate this interpretation.

First, rising sea levels are creating changing landscapes of inequality – drawing new world maps whose key lines are not traditional boundaries between nation-states and social classes but elevations above sea or river. This is a totally different way of conceptualizing the world and our chances of survival within it (chapter 4).

Second, climate change produces a basic sense of ethical and existential violation that creates new norms, laws, markets, technologies, understandings of the nation and the state, urban forms, and international cooperations.

Third, the Cosmopolitan Turn 2.0 is unfolding not in the thinking about the world and the withering of national doctrines but firstly and mainly in the reality of everyday practices and activities. The insight that no nation-state can cope alone with the global risk of climate change has become common sense. From here arises the recognition of the fact that the principle of national sovereignty, independence and autonomy is an obstacle to survival of humankind, and that the 'Declaration of Independence' has to be metamorphosed into the 'Declaration of Interdependence': cooperate or die!

Consequently, 'methodological nationalism', the teachings of turning the world around the nation, has to be substituted by 'methodological cosmopolitanism', the teachings of the turning of the nation around the 'world at risk'.

If we consider how the issue of climate change fits into the current perspective in politics and the social sciences, we can see the limitations of 'methodological nationalism'. We frame almost every issue, whether it relates to class, conflict or politics, in the context of nation-states organized within the international sphere. But, when we look at the world from the perspective of climate change, this framing doesn't fit. A new power structure is embedded within the logic of global climate risk. When we talk about risk, we have to relate it to decision-making and decision-makers, and we have to make a fundamental distinction between those who generate risk and those who are affected by it. In the case of climate change, these groups are completely different. Those who make decisions are not accountable from the perspective of those affected by risks, and those affected have no real way of participating in the decision-making process. It's an imperialistic structure; the decision-making process and its consequences are attributed to completely different groups.

We can observe this structure only when we step outside of a nation-state perspective and take a cosmopolitan perspective, where the unit of research is a community of risk that includes what is excluded in the national perspective: the decision-makers and the consequences of their decisions for others across space and time.

2   Metamorphosis is about a New Way of Generating Norms

Climate change is creating existential moments of decision. This happens unintended, unseen, unwanted and is neither goal-oriented nor ideologically driven. The literature on climate change has become a supermarket for apocalyptic scenarios. Instead, the focus should be on what is now emerging – future structures, norms and new beginnings.

Metamorphosis is about a new way of generating critical norms in the age of global risks. Legal scholars and standard sociology think about violation only if there is a norm. But, with global risks, a new global horizon is emerging from the experience of the past and expectation of future catastrophes. The sequence is turned upside down – the violation comes *before* the norm. The norm arises from the public reflection on the horror produced by the victory of modernity. A brief look at the history of world risk society illustrates this metamorphosis. Before Hiroshima, no one understood the power of nuclear weapons; but, afterwards, the sense of violation created a strong normative and political momentum: 'Never again Hiroshima!' Violations of human existence such as that at Hiroshima induce anthropological shocks and social catharsis, challenging and changing the order of things from within (chapter 7).

'Never again Holocaust!' This metamorphosis decouples our normative horizons from existing national norms and laws by introducing the notion of 'crimes against humanity'. I am referring here to something profound. A basic principle of national law was that an act could not be judged in hindsight under a law that did not exist at the time the act was committed. So, while it was legal under Nazi law to kill Jews, in hindsight it became a crime against humanity. It was not simply a law that changed, but our social horizons – our very being in the world. And the change took place in an unexpected way that enforced the global gravitational pull of social and political action (a human rights regime). This is exactly what I mean by metamorphosis: what was utterly unthinkable yesterday is possible and real today, creating a cosmopolitan frame of reference.

Given the reality of cosmopolitization, the rebirth of the national outlook is paradoxical. It characterizes the schizophrenic structure of the *Zeitgeist*. It governs thinking, while social activities, if they want to be successful, open themselves up to the cosmopolitan field of action. And it is the national outlook in public and academic discourse that blinds us to the alternatives for climate change action that we see from a cosmopolitan point of view.

3   Climate Change: Nature and Society Combined

In the case of climate change as a moment of metamorphosis, there is a coalescence of nature, society and politics. Therefore the narrative of the risk society is in itself a narrative of metamorphosis of the world. It is a narrative about a human condition without precedent. It provides a way of speaking of the physical world and of its risks that brought in a striking array of new topics. It enables people to speak of things – indeed, in a way to see things that they had been trying to speak of and to see – but where the concepts had been chronically lacking. Metamorphosis in the terms of risk society means the end of the distinction between nature and society; I quote from *Risk Society* (1992: 80):

That means that nature can no longer be understood *outside of* society, or society *outside of* nature. The social theories of the nineteenth century (and also their modified versions in the twentieth century) understood nature as something given, ascribed, to be subdued, and therefore always as something opposing us, alien to us, as *non*-society. These imputations have been nullified by the industrialization process itself, *historically falsified*, one could say. At the end of the twentieth century, nature is *neither* given, *nor* ascribed, but has instead become a historical product, the *interior* furnishings of the civilizational world, destroyed or endangered in the natural conditions of its reproduction. But that means that the destruction of nature, integrated into the universal circulation of industrial production, ceases to be 'mere' destruction of nature and becomes an integral component of the social, political and economic dynamic. The unseen side effect of the societalization [*Vergesellschaftung*] of nature is the *societalization of the destruction and threats to nature*, their transformation into economic, social and political contradictions and conflicts. Violations of the natural conditions of life turn into global social, economic and medical threats to people – with completely new sorts of challenges to the social and political institutions of highly industrialized global society.

An example of this can be seen in how industry internalizes and revises climate costs. Transnational companies – such as Coca-Cola – have always been focused more on their economic bottom line than on global warming. But when the company loses a lucrative operating licence, for example, in India, because of a serious water shortage, the perceptions and priorities begin to change. Today, after a decade of increasing damage to Coca-Cola's balance sheet as global droughts dried up the water required to produce its soda, the company has accepted that climate change is an economically disruptive force:

> We believe that climate change, caused by man-made green-house gas emissions, is the greatest threat to our planet. There is an urgent need for a step-change to achieve not only the significant emission reduction targets we have set but also a low-carbon future. To that point, we have to look beyond our own operations and take responsibility for the whole product value chain. (See www.eurotrib.com/story/2014/1/25/12338/0822)

More droughts, more unpredictable variability, and 100-year floods that take place every couple of years are disrupting the company's supply of sugar cane and sugar beets, as well as citrus for its fruit juices. 'When we look at our most central ingredients, we see those events as threats', said one of its responsible managers.

This reflects a new awareness among American and European business leaders, and mainstream economists as well, who see global warming as a force that contributes to lower gross domestic products, higher food and commodity costs, broken supply chains and increased financial risk. Their position is at striking odds with the long-standing argument, advanced by economists and managers of transnational companies, that policies to curb carbon emission are more economically harmful than the impact of climate change itself. Internalizing the destruction of 'nature', an economic study on the production and financial risks associated with climate change demonstrates how, in the age of climate change, industrial business is becoming 'risky business'. Thus industry is awakening to the effects of climate change and their real costs. Thus climate risks, or the Anthropocene (Crutzen 2006) – a new geological era in earth's history, in which humans are the defining ecological force – enters into the realm of business and economics. This makes the causes, consequences and responses to global climate change fundamentally social and political in nature. Here metamorphosis means that climate change is about humans shaping the direction of planetary and social evolution – not by intention but by the politics of side effects or the politics of normalized damage.

4   Global Risk Comes as a Threat and Brings Hope

Global risk is not global catastrophe. It is the anticipation of catastrophe. It implies that it is high time for us to act – to drag people out of their routines and pull politicians from the 'constraints' that allegedly surround them. Global risk is the day-to-day sense of insecurity that we can no longer accept. It opens our eyes and raises our hopes. This encouragement is its paradox. There is a certain affinity between the theory of world risk society and Ernst Bloch's (1995) principle of hope. World risk society is always a political category; it creates new kinds and lines of conflict and liberates politics from existing rules and institutional shackles.

This again is what I mean by metamorphosis. Climate change might in fact be used as an antidote to war. We are undergoing a transition from the threats emanating from the logic of war to those arising from the logic of global risk. In the case of war, we find rearmament, resistance to enemies or their subjugation; in the case of risk, we see cross-border conflicts, but also cross-border cooperation to avert catastrophe – this is what I refer to as cosmopolitization. Thus life and survival within the horizon of global risk follow a logic that is diametrically opposed to war. In this situation it is rational to overcome the us–them opposition and to acknowledge the other as a partner rather than as an enemy to be destroyed. The logic of risk directs its gaze towards the explosion of plurality in the world, which the friend–foe gaze denies. World risk society opens up a moral space that might (though by no means necessarily will) give birth to a civil culture of responsibility that transcends old antagonisms and creates new alliances as well as new lines of conflict.

Global risk has two sides: the traumatic vulnerability of all and the resulting responsibility for all, including one's own survival. It forces us to remind ourselves of the ways in which the human race jeopardizes its own existence. Consciousness of humanity thus acts as a fixed point. The risk of climate change generates an *Umwertung der Werte* (a revaluation of values) (Nietzsche), turning the system of value orientation upside down – e.g., from postmodern cultural relativism to a historical new fixed star by which to mobilize solidarities and actions. This is the case because the global climate risk contains a sort of navigation system in the otherwise storm-tossed seas of cultural relativism.

Whoever speaks of mankind is not cheating (as Pierre-Joseph Proudhon and Carl Schmitt put it) but is forced to save the others in order to save her- or himself. In world risk society, cooperation between foes is not about self-sacrifice but about self-interest, self-survival. It is a kind of egoistic cosmopolitanism or cosmopolitan egoism. We have to distinguish between a neoliberal form of self-interest and the self-interest of humanity.

But metamorphosis is not a direct line to a cosmopolitan future, in a normative political sense. In fact, the opposite is the case: metamorphosis is highly ambivalent. While victims of climate change, such as small island states, are being repositioned on the global map, there might still be new imperialistic orders emerging. The danger of 'climate colonialism' is very real. We have to take a cosmopolitan outlook to make these vulnerable situations visible, tangible, and ask what consequences for thought and action they have in the West. How can we give them a voice in 'our' political processes? This would indeed require a redefinition of national interest.

5   World Cities are Rising as Cosmopolitan Actors

We also are undergoing a metamorphosis of the landscape of global actors through which nation-states are becoming cosmopolitized. On the one hand, nation-states are realizing that there are no national answers to global problems, even facilitating networks of global cities as cosmopolitan actors. On the other hand, national institutions are still subject to and products of the imagination of sovereignty.

Normative cosmopolitan expectations thus produce both cosmopolitan nations and renationalizing nations. Renationalizing nation-states are paralysing cosmopolitan cooperation; international conferences fail. So, to find answers to climate change, we should look not only to the *United Nations* but also to the *United Cities*.

Social movements are important for setting the cosmopolitan frame, but they do not create collectively binding decisions. For this, there is the nation-state, with its monopoly on law-making. But the influence of the nation-state is eroding. World cities are becoming more important spaces for setting collectively binding decisions. Why? In cities, climate change produces visible effects; climate change incentivizes innovation; cooperation and competition transgress borders; and political response to climate change serves as a local resource for political legitimation and power.

A new power structure is emerging; it is composed of urban professionals in world cities – urban transnational classes with varying historical backgrounds. Cities are being legally redefined as transnational actors, organized voices of transnational politics. Even Zurich is a mini New York; it is not one city but many world cities in one, with a strong red–green coalition urban government and few chances for conservatives to regain power.

There are also basic contradictions. Urbanization used to be defined in opposition to nature. Nowadays, it is the other way around: 'green urbanism' is everywhere; 'sustainability' has become normalized; everything is now about greening.

But these kinds of deconstructions are legitimizing the new normative horizon of cosmopolitan expectations. World cities are creating a new world of inclusiveness, where the potential to transform the law is growing. Making this new potential visible is what my theory of metamorphosis is all about (chapter 12).

6   Pascal, God and Climate Change

Let's do a thought experiment: climate change scepticism can be a strong position. What then is the counter-argument? My counter-argument refers to the French philosopher Blaise Pascal and his pragmatic 'proof' of God. Pascal argued that either God exists or he doesn't. I don't know. But I have to choose God because, if God exists, I win; if he doesn't, I don't lose anything.

Let's compare the belief in God with the belief in human-made climate change. Like Pascal, we do not know if climate change is 'real'. Despite substantial evidence, a basic uncertainty remains. We need to accept that it is impossible to know if a natural catastrophe is actually the consequence of human-made climate change. This uncertainty creates a critical political moment of decision.

There are two scenarios. First, we deny climate change, which would mean that every catastrophe highlights the irresponsibility of deniers. Second, we accept that climate change is real, take responsibility, and confront the overwhelming scale of necessary moral and political change. As in Pascal's case, there are good, practical reasons, even for the deniers, to accept that climate change is real. Climate change may change the world for the better.

Seen as a global risk to all civilization, climate change could be made into an antidote to war. It induces the necessity to overcome neoliberalism, to perceive and to practise new forms of transnational responsibility; it puts the problem of cosmopolitan justice on the agenda of international politics; it creates informal and formal cooperation patterns between countries and governments that otherwise ignore each other or even consider each other enemies. It makes economic and public actors accountable and responsible – even those who do not want to be accountable and responsible. It opens up new world markets, new innovation patterns, and the consequence is that deniers are losers. It changes lifestyles and consumption patterns; it reveals a strong source of future-oriented meanings, in everyday life and for legitimation of political action (reforms or even revolutions). Finally, it induces new forms of understanding and caring for nature. All of this happens under the surface of the mantra of disappointments and disillusionments at the *travelling circus* of one climate conference after the other.

From this perspective, climate change means first the metamorphosis of politics and society that has to be discovered and closely analysed through the social science of methodological cosmopolitanism. This is not to say that there is an easy solution to climate change. Nor is it to say that the positive side effects of negative side effects automatically create a better world (chapter 7). And it is not even to say that the active, sub-political and political metamorphosis is fast enough to counter the galloping process of climate catastrophes that might throw the entire planet into droughts, floods, chaos, starvation and bloody conflicts. But, ultimately, catastrophe too would be a metamorphosis – the worst mode of metamorphosis.

4

Theorizing Metamorphosis

1 The Return of Social History

Someone who wants to explore how certain facets of the metamorphosis of the world appear or, alternatively, fail to appear in certain contexts and themes must raise the question of the return of social history. What is special about the return of social history is that, in the light of metamorphosis, it cannot be demonstrated in terms of intentions, ideologies, utopias or political programmes and conflicts, class struggles, refugee movements or wars. Rather, it slinks in, as it were, through the back door of side effects. The interpenetration of side effects and global historical change is the joke and punchline of the argument.

The reflexivity of second modernity is a result of the fact that societies are now confronted with the undesirable side effects of their own modernizing dynamics, which they have often consciously accepted as collateral damage. It is not poverty, but wealth, not crisis, but economic growth, coupled with the suppression of side effects, which are driving the side-effects metamorphosis of modern society. This is not abolished but, rather, accelerated through inaction. It does not come from the centres of politics but from the laboratories of technology, science and business.

This metamorphosis via side effects prevails because it is expressly *not* made into a topic for an election, hence not through democratically tamed politics but in virtue of the power of concealed side effects. In this way, nationally organized industrial society is becoming metamorphosed into an unknown world risk society.

To adapt John Dewey's argument in his book *The Public and its Problems* (1954), world risk society is a social formation in which the accepted, accumulated side effects of billions of habitual actions have rendered the existing social and political institutional arrangements obsolete. In the metamorphosis that becomes thematic with world risk society, the side effects of past action, which have turned into the main effects, have permeated society as a whole in such a way that they are creating a growing awareness that the narrative of the controllability of the world has become fictional (to repeat, in varying degrees in different contexts, cultures and corners of the world). That this idea is also productive for historiography has been expounded by Benjamin Steiner (2015: 33–4):

> In addition, however, accepted side effects provide historiography with a heuristic model for representing historical change. Only at first sight do side effects appear as a surprising occurrence in history. If subversive and erosive follow-on problems are problematized within a crisis discourse, this does not mean that their occurrence was not foreseen. As will be shown in the following historical examples of epochal breaks, this is how the rough examination of a crisis discourse reveals that side effects are often blamed for the crisis. It is criticized that slow movement within a discourse associated – or not associated – with certain actors gives rise to a contradiction within our own collective social self-understanding. From the perspective of a crisis discourse, therefore, follow-on problems or side effects appear to be unintentional, or they are causally attributed to the intention of a generally minority and, therefore, relatively ineffectual, counter-discourse. It is always disconcerting in this context that the problems were not only unintended and find a home outside the mainstream in the 'underground', but that they

are also seen to be inherent in the main discourse and ultimately to have been already implicit in it. And, thus, we have to sharpen our vision by using the side-effects spectacles and focus them on the transitions from main discourses to secondary discourses.

When it comes to contemporary sociology, one has to say that the mainstream theories in sociology, but probably also in political science, and the research strategies based upon them are not able to register and acknowledge the return of social history. This is because the major theories of a Foucault, a Bourdieu or a Luhmann, as well as rational choice theories, notwithstanding all their differences, have one thing in common: they focus on the reproduction, and not on the transformation, let alone the metamorphosis, of social and political systems.

Grasping these transformations, however, requires a fundamental break with the dominant metaphysics of social reproduction, which always shows the circular re-emergence of the same basic patterns and dualisms of modernity. Such a break, however, which acknowledges the re-emergence of historicity, represents an epistemological and political threat, in that it challenges the established scientific disciplines and their various monopolies on expert authority. This is visible, for instance, in the way presumptions of the reproduction of socio-political order are built into dominant constructions of globality, including in macro-economic forecasts and techno-scientific constructions of the global climate (cf. Guyer 2007; Szerszynski 2010). Framed by the metaphysics of reproduction, such globalities may be learned, exported, and used as a common model for integrating and domesticating politics. Since the future is conceptualized as part of the experience of the past, there is no basic disconnect, only a matter of linear extensions. It is a model close to timeless eternity: present society dominates and colonizes the future, thus rendering it controllable.

To sociology, breaking with the reproduction of the social order and theorizing (cosmopolitan) metamorphosis implies its own set of epistemological and methodological difficulties. In First Modernity, there exists an elective affinity between orthodox appeals to the reproduction of social structures and the practice and authority of empirical sociology: the metaphysics of reproduction allows for the establishment of social laws and regularities, enabling sociologists to make prognoses, to do comparative studies, and so on. In Second Modernity, the situation of sociologists is akin to what Tocqueville said of the 'human spirit': if modernity breaks with continuity, since the past stops to throw its light on the future, the human spirit (i.e., the sociologist) is lost in darkness! When taking historicity seriously, then, sociologists find themselves in a difficult situation, since they can no longer use the past or the present to talk about the future; from now on, they have to concentrate on the future itself, without the security blanket of the past. Cosmopolitan sociology, in short, must reorient itself towards an unknown and unknowable future, made present in the temporal horizons of global risk.

2　Forms of Historical Change: the Axial Age, Revolution, Metamorphosis of the World and Colonial Transformation

The sociological concept of the metamorphosis of the world refers to an unprecedented historical form of global change that involves two levels, the macro-level of the world and the micro-level of everyday human life. Their specificity becomes most apparent when one compares them – in a very simplified and sketchy way – with three well-known forms of historical change: with the so-called Axial Age, with the French Revolution and with colonial transformation.

Axial Age

A fundamental form of historical change occurred as the overthrow of the religious world pictures, as thematized in the discussion of the so-called Axial Age (Karl Jaspers, Shmuel Eisenstadt). These were revolutions in worldviews which – this is the crucial point here – occurred exclusively in, and remained confined to, the self-contained parallel universe proper to theology. They affected only the 'superstructure', without extending to society. They had little influence on structures of rule and class relations, on the gender hierarchy and the economy, and hence on people's everyday lives.

In this revolution in worldviews, which began in the religious cultures of the Axial Age (400 BC), fundamental tensions broke out between the transcendental religious order and the temporal order. Whereas in earlier eras the belief prevailed that the afterlife and this world form a unity, the Axial Age marked the beginning of theological and philosophical disputes between *spiritual elites* who sought to shape the world in accordance with their transcendental visions. They elaborated different visions of a divine moral order which legitimizes the temporal social and political order. Since the transcendental order was conceived as conclusively justified and foundational, it was the yardstick by which the temporal order was measured. Even the ruler was now confronted with a superior authority.*

Over the course of the following centuries, these visions of a transcendental order became repeatedly embroiled in tensions with emerging intellectual and philosophical currents and new scientific discoveries and theories. The representatives of Christian dogma responded to the resulting challenge

---

* In a sense, one can say, following Shmuel Eisenstadt, that the distinction between the this-worldly (temporal) and other-worldly (transcendental) orders was first introduced with the Axial Age. If the other-worldly and this-worldly orders are ultimately represented as indistinguishable, there cannot be a question of the other-worldly, transcendental order being superior, nor can there be any problem of the legitimacy of the temporal order. 'By contrast, in the Axial Age civilizations the perception of a sharp disjunction between the mundane and transmundane worlds developed. There was a concomitant stress on the existence of a higher transcendental moral or metaphysical order which is beyond any given this- or other-worldly reality' (Eisenstadt 1986: 3).

by trying to integrate these currents, but without renouncing the dominance of the closed theological worldview.

The main concern was to deflect attacks by the rapidly advancing sciences of nature. The image of the world that had been preached until then was turned on its head by the revolutionary discoveries of a Galileo, for example. The effect was abrupt. Until then, theology had taught that the earth was a disc and that anyone who strayed too far from the centre would fall off. Now it had to concede that the earth is a sphere, but at the same time it tried to preserve the theological image of the world and, with it, the ecclesiastical temporal power structure. Similarly, theologians later faced the difficult task of combining the theological view of the world with the knowledge that (contrary to appearances) the sun does not revolve around the earth, but the reverse – that the earth revolves around the sun and that everything earthly and human is only a minute segment of an infinite space.

This form of historical change involves transformations that essentially remain confined to the sphere of influence of theology and the discourses of intellectual elites and do not reach society and people's everyday lives – or, if they do, they do so only very indirectly. Even where social and political changes are envisaged, these remain tied to the interests of intellectual elites, however comprehensively they may be formulated. The economic basis and the existing relations of domination were not affected, and a redistribution of the means of production was not sought. Moreover, the potentially revolutionary discourse about the equality of human beings was politically tamed by a form of political and theological metaphysics which postulated that existing inequalities – in particular, the fact that slaves and women had no rights – were prescribed by nature.

Revolution

A second form of epochal change is revolution, the paradigmatic example of which is the French Revolution. Here the transcendence of the theological worldviews is given a political turn. The idea of equality, which is implicit in the Christian worldview, is transformed into the revolutionary ideas and utopias that blow apart the feudal power relations. The characteristic feature is that revolution is a matter not just of changes at the level of theological and philosophical worldviews but of overcoming the assumed naturalness of the social and political order and of opposing to it the utopia of the malleability of politics and society. As a result, everyday life was now also gripped by the dynamics of historical change.

These revolutionary changes manifested themselves in three dimensions: here the *idea of equality* that convulsed the social order from the very top to the very bottom broke new ground.

Moreover, the concepts of reason and rationality stemming from the Enlightenment combined to produce a *radical critique of religion*. The core idea of the Enlightenment amounted to an attack on religion because it fundamentally questioned the legitimacy of the religious basis of value, the justification of the temporal by the other-worldly order. The modern utopias of liberty, equality and fraternity no longer needed a god as the source of their legitimacy.

Against this background, the *idea of nationalism* also spread. With its triumphal march from the nineteenth century onwards, first in Europe and later everywhere, this idea structured the world in accordance with the key distinction between national and international and imposed this order – not only in the political sphere but also in the sciences (history, political economy, sociology, international relations, anthropology, ethnography, and so forth).

Revolution is not a one-way concept. The French Revolution was succeeded first by the Marxist Russian Revolution and then by the racist National Socialist revolution in Germany (and other European countries). This culminated in the catastrophes of the First and the Second World Wars, the Holocaust, the Gulag, and the bipolar global order of the Cold War.

From the cosmopolitan perspective, nationalism is particularly toxic not only because of the overt justification it gives to global wars and global inequalities. It is dangerous on account of its cognitive status: nationalism defines and ossifies our political and social scientific frameworks and our most basic categories of thought and knowledge. Nationalism as an ideology thus limits not only what we can imagine and wish for but, more importantly, what we know and how we conceive of reality. The most basic categories are indeed captive to the national order: citizen, family, class, democracy, politics, state, etc. – all are nationally defined. Our legal and administrative systems define them, and these definitions are amplified by the social sciences through methodological nationalism (Beck 2000, 2006; Wimmer and Glick Schiller 2002, 2003).

The concept of the metamorphosis of the world is much more than a social and political theory, a utopia (or dystopia) – it is the reality of our times. I turn the argument that 'cosmopolitization' is an unrealistic ideology on its head, arguing that, at the beginning of the twenty-first century, it is the proponents of the national who are the idealists. They view reality through the obsolete lens of the nation-state and thus cannot see the metamorphosis of the world. The sociology of metamorphosis, therefore, is the critical theory of our times, since it challenges the most profound truths which we hold dear: the truths of the nation.

Metamorphosis

There are two historical preconditions that enabled the historically unprecedented metamorphosis of the world finally to begin – the collapse of imperialism and the collapse of the Soviet Union, and with it the bipolar global order. It happened not least as a side effect of what is so casually referred to as 'globalization'. The colonial transformations were transcontinental but not global in a strict sense. Unlike the French Revolution, the metamorphosis of the world is not confined to the political centre of the regime. Rather, it is everything simultaneously: local, regional, national and global – albeit in the contemporaneity of the non-contemporaneous. Unlike revolution, it affects not only a political regime but also the understanding, the concept of the political and of society itself. It is not just a temporally, spatially and socially limited exception (like revolution or war) but progresses ever further, and even escalates with the escalation of risk capitalism. It is not intentional, programmatic and ideological, and it is not undermined but, rather, kept going by political inaction. It does not emerge from the centres of democratically legitimized politics but instead proceeds – as a socially and legally constructed 'side effect' – from the profit calculations of economics, from the laboratories of technology and science.

Thus revolutionary consciousness is also foreign to the metamorphosis of the world. That a metamorphosis is taking place which is subverting not just the national order but, imperceptibly and unintentionally, even the world order must first be gradually revealed in politics, science and everyday life – against the insistence of the dominant view which understands the world in categories of immanent social change or linear transformation.

This all-encompassing, non-intentional, non-ideological metamorphosis that takes hold of people's daily lives is occurring almost inexorably, with an *enormous acceleration* that constantly outstrips existing possibilities of thought and action. While the conflicts over revolutions in worldviews lasted decades, even centuries, while the effects of the French Revolution extended over the past 200 years (and are still continuing), the metamorphosis of the world is taking place in world seconds with a speed that is nothing short of *inconceivable*, and as a result it is overrunning and overwhelming not just people but also institutions. That is why the metamorphosis that is at present occurring before our eyes is almost beyond the reach of the conceptualization of social theory. And that is also why many people now have the feeling that the world is unhinged.

This acceleration becomes especially apparent at the level of language. Under the pressure of metamorphosis, many key political and social concepts are becoming anachronistic and are even becoming so hollowed out from within that they no longer disclose anything. Whether it is the opposition between left and right in politics, distinctions such as those between nationals and foreigners, nature and society, First and Third World, centre and periphery – everywhere we find deflated linguistic formulas, broken coordinates and hollowed-out institutions. Familiar concepts are becoming memory traces of a bygone era. At the same time they are the writings on the wall that announce the metamorphosis of the world.

Colonial transformation

Colonial transformation is an early stage of a kind of imperialistic globalization before globalization. Colonialism is in fact as old as civilization. It is an integral part of the history of all civilizations and religions in both the East and the West. Yet, driven by the idea of universal Christianity, Western colonialism came closer to the goal of global dominance. Colonial power involved an unimaginable degree of violence and cruelty, which was legitimized through the ideology that the 'non-believers' needed to be converted for the sake of their own souls. Any opposition, such as the worldview and beliefs of the colonized, was destroyed by the combination of conquest and missionary work. Columbus captured this in the principle that 'those who are not already Christians can only be slaves'. This is why Western colonialism needs to be understood as a hierarchical entanglement between centre and periphery. The stability of colonial power was grounded not least in the fact that a notion of inferiority and primitivism was imprinted on the colonized through violence and, in fact, became part of their self-understanding.

As with revolution, the model of historical transformation of colonial power is characterized by intention, goal, religion, politics, violence, dominance and ideology.

The birth of the European nation-states would not have been possible without the exploitation of humans and material resources within the colonized territories. The colonies constituted 'laboratories of the future' in which what would subsequently be implemented in the European nation-states was tested. Here, we can see the beginnings of cosmopolitization, which constitute a significant part of the metamorphosis at the beginning of the twenty-first century. This is not to equate colonialism with an early version of cosmopolitization. Although both forms of entanglement are asymmetrical, it is meaningful to speak of cosmopolitization only if the asymmetrical entanglements are perceived in the light of anticipated equality.

But there is still a big problem: Are we really witnessing a 'metamorphosis' from neocolonialism to cosmopolitization? What kind of processes and steps do we have to distinguish, and what criteria do we have in order to answer this question? Or, even more fundamentally: What does 'enforced inclusion of global others', the definition of cosmopolitization, exactly mean? Who is forced by whom to do what?*

Explaining the 'metamorphosis of postcolonialism' actually means to distinguish carefully between colonization and cosmopolitization.

Metamorphosis starts with the distinction between dependency (theory) and cosmopolitization (theory). Both describe global forms of historical transcontinental inequalities and asymmetrical power relations. But their social and political quality is changing because cosmopolitization creates *normative horizons of equality and justice*, thereby generating pressure for inclusive change on the existing structures and institutions of global inequality and power (chapter 6). This first process of metamorphosis relates not necessarily to a decrease of asymmetries (there could even be an increase of global inequalities) but to the implementation of global norms of equality. This is happening through

---

* These problems are not to be confused with the discussion on the relationship between normative philosophical cosmopolitanism and colonialism (see Köhler 2006, etc.).

the human rights regime, its institutionalization and the global advocacy around it. They transform existing global hierarchies from (what was perceived by the colonizers as) 'naturally given ("goods")' into 'political bads', violating the normative order of the world.

The second process of metamorphosis refers to global risks intensifying and (trans-)shaping worldwide social relations, however uneven and sporadic, thereby creating moments of *shared fate*. The metamorphosis produced by global risks transfigures uni-directional imperialism into the global spread of manufactured uncertainties – a shared problem that cannot be solved nationally or by referring to the old dualism of the 'colonial' and the 'postcolonial'.

Both – the normative horizon of equality and the shared problem of manufactured uncertainty – produce reflexivity: the 'entangled histories' (Randeria) produced by colonialism are being remembered and redefined in the light of endangered futures.

Ana Maria Vara (2015) argues that the metamorphosis of neocolonialism into cosmopolitization depends fundamentally on the structures and resources of power in a very specific way. There have to be strong elements of *reversed* dependencies. This means *Umwertung der Werte*: reverse valuation of naturally given asymmetries into 'political bads' is a necessary but not a sufficient condition. Additionally there has to be what one might call the 'emancipation of power'. This implies that the ex-colonizer depends on the growing power of the ex-colonized. One can argue that the process of metamorphosis further depends on facts demonstrating that the (post)colonially excluded are included in the negotiations on world affairs because of the emancipation of power. All this implies new landscapes of danger and hope.

> On the national layer, regarding electric cars, Bolivia, Chile and Argentina are expected to provide as usual the natural resource, lithium; while Japan, Germany, or South Korea are expected to industrialize it, and to provide the technology, the batteries, and/or the cars – and, in turn, to buy the cars. What does it imply? Where is metamorphosis here? Bolivia,

Chile, and Argentina are currently working on and negotiating from the position of suddenly empowered nation-states in a new geopolitical world. Maybe cosmopolitization has to do with the *current power* for negotiating the terms of the relationship, and a *future horizon of some kind of symmetrization* of the relationship. Let's imagine the nationals of these three South American countries as saying: 'We are not equal. But we have the right *to aspire to be equals*. And to be recognized as having this right.' (Vara 2015: 102)

From this, Vara argues that cosmopolitization, seen as creating a normative horizon, implies the politics of possibility to transform the power relationship: 'not reversing it, not inversing it, but something else, as metamorphosis implies' (ibid.). How socially prevalent is metamorphosis? Metamorphosis, in principle, is unfinished, unfinishable, open-ended and – this is important to notice – irreversible, but it may still be instrumentalized for imperialistic purposes.

3   Risk Society as Agent of Metamorphosis

Those who bemoan the current Eurocentrism of philosophy, geography, sociology, the feminist movement, environmental criticism, or even politics in general cannot expect to inspire curiosity and command people's attention. The surprising thing is, instead, how familiar and normal this decentralizing critical act has become. On the other hand, however, this 'yawn' with which the critique of Eurocentrism is greeted makes clear that the point of the criticism – namely, the call to establish a worldview that is not centred on Europe – has long since arrived and been accepted, even if people still puzzle and argue over what it means and whether it has any practical effects. Also, the now commonplace critique of the imperialism of the West lives off the demand to overcome this and to establish a 'good world' in which Western imperialism has been overcome.

Just as the religious worldview was eroded by scientific criticism of religion and was replaced by the modern worldview in which scientific rationality holds sway, today the dismantling of all direct and indirect forms of privilege is demanded in the name of an even more rational rationalism, whether the target of the critique be the rich, the white peoples of the northern hemisphere, or even the Westphalian system of states. The contours of the new structures of plausibility and rationality become apparent not in what is established, but in what is criticized. This is also, or even especially, the case when these new rational standards of equality and justice are *not* realized.

Technological deterministic optimism

It is not only world risk society that is transforming the world (as I will show in greater detail later). There is a different mode of metamorphosis of the world. This is the new technological deterministic optimism, which is shaped by a healthy ignorance of the impossible. The modern view of the world was based on what was called 'faith in progress' – namely, the transfer of the religious belief in salvation to the earthly, secular productive forces of technology and science. Here, too, belief means trust in the invisible, in this case in the capabilities of human beings still to be developed and of their institutions to solve the problems of existence with increasing precision and efficiency. We still behave largely as if this were the ultimate truth. This is reflected in turn in the global statistics and reports that at least impugn, and perhaps even refute, this belief. People bemoan the high levels of illiteracy, of childhood illnesses, of the new civilizational diseases, of overpopulation (or of underpopulation), of traffic fatalities, of environmental destruction, of economic stagnation, of human rights violations, and so forth. All this is deplored, but it is simultaneously demanded and established as a benchmark for 'good action' – specifically, this is the point, for the 'world' and 'humanity'.

As Joshua J. Yates (2009) impressively demonstrates, this mixture, which calls for the cosmopolitan normative framework in a kind of self-fulfilling prophecy, is reflected in different maps of the world that thematize the corresponding distributions in graphic form. In this context, one can think of many global maps of failure and injustice. It is a matter of the radically unequal distribution and perception of opportunities for consumption in the world, or of how HIV-infected persons are spread across the globe (where the powerful continental figures of Europe, America or Russia and Asia are shrinking, while the various African regions are expanding into unrecognizable, cumbersome structures that fill most of the world map). Something similar can be seen in the case of the world map of poverty. In contrast, on the global map of gross domestic product, the Western countries and the Asian Tiger countries feature as mammoth figures dwarfing continental midgets such as Africa and South America. A global map also recounts how earnings and happiness are connected (or not) across the world. And those who do not believe any of this will find evidence for their rejection in the depiction of how the values for GDP and GPI (genuine progress indicator) have been drifting apart worldwide since 1950.

How the world is becoming smaller is shown not least by how overseas tourism and risk zones overlap and interpenetrate. Europeans are flocking to the warm regions; but many of these destinations count as dangerous, whether because of deadly diseases, natural disasters, increasing poverty or wars. The result is a kind of liquid belonging subject to revocation in which hopes, fears and disappointments intermingle. All of this can be read as initial pointers for a sociological study of the metaphysical metamorphosis of the modern world along the path leading to a different world.

Accordingly, a gulf has opened up. The classical worldview of the modern faith in progress still guides action – the belief in the redemptive power of techno-science, the idea of limitless progress, the inexhaustibility of natural resources, the belief in infinite economic growth and the political supremacy of the nation-state. The theory of risk society has confronted this belief with its theoretical fragility and inadequacy in view of the scenarios of catastrophic potentials and uncertainties currently unfolding, which are precisely the result of the triumphs of progress. But technological and moral weapons are being developed against this, and not just in Silicon Valley. They assume the form of a super-optimism of progress that liberates the world from all of the evils produced by modernity – preventing cancer, prolonging life, overcoming poverty, stopping climate change, abolishing illiteracy, and so forth – the new crusaders of the technological faith in progress promise all of this. 'People in Silicon Valley are convinced that they are not only delivering products, but revolutions', according to the Stanford University futurologist Paul Saffo. Here people are working on 'moon shots' – that is, on really big things that will utterly change the world. The twinkle in the eyes of the new technological world changers is disconcerting. For here the key argument of the theory of world risk society is being flouted. This involves three steps: first, the paradox that those who ignore the destructive side effects of the triumphs of modernization (the belief in progress) accelerate the latent process of destruction and intensify and universalize it. The social and political threats must be clearly distinguished from this physical destruction and threat. The subject of the theory of risk society is not (only) physical destruction and global risks but their social, political and institutional consequences. Whether the world ends or not is, sociologically speaking, completely uninteresting. In contrast, of major sociological importance is the idea, as developed in the concept of world risk society, that the environmental side effects of industrial capitalism bring with them a socially transformative power – and one of metaphysical proportions. In other words, risk society is the product of the metamorphosis that has become the productive force and the agent of the metamorphosis of the world.

The second key argument is that the destructive consequences of industrial production cannot be externalized forever. Quite to the contrary, it is precisely the super-faith in progress which – by stubbornly ignoring, belittling, denying the existence of risks – gives birth to, magnifies and globalizes new global risks of unknown proportions. To give an example, the United States can opt out of the Kyoto process to place limits on greenhouse gases, but sooner or later it will have to face the consequences, whether in the form of catastrophic climate effects (hurricanes, etc.), in political reactions by other affected countries, populations, continents and states, or as a global domestic political conflict.

A third major aspect of metamorphosis concerns the influence of global risks on the awareness or becoming aware (of metamorphosis) itself. This is a matter, on the one hand, of 'reflexivity' (self-confrontation) and, on the other, of reflection (knowledge, global discourses). The environmental conflict is taking place *not* in the 'environment' itself but in institutions, political parties, trade unions and global corporations, within and between governments and international organizations, or at the breakfast table, where everything turns on the legitimacy of lifestyles, of breakfast and of consumption. Here, the viewpoint that is crucial for metamorphosis repeats itself: the complaint and criticism that in the end nothing is happening, that everything is remaining as it is, is precisely the paradoxical way in which the radical change in horizon is occurring, in which the new fixed stars are being established that bear the proud names of 'world', 'humanity' and 'planet'.

However, this is precisely not a matter of 'thin air' (as is often also lamented). Rather, this is how globalized forms and spaces of action arise – that is, globally available models of protest and resistance. These belong, as it were, to the aggregate of the cosmopolitized field of action. Correspondingly, they can be activated by action groups from civil society and non-civil society movements to bring about global change.

Many are under the impression that, with the implosion of Eastern European state socialism, any form of social criticism 'that gets at the root of the problems' has become impossible. In fact, the opposite is the case. A cosmopolitized world marked by a high level of reflexivity, in which the problematization of all social relations has become something to be taken for granted and in which the scope for cosmopolitan action is increasing, actually stimulates political and scientific criticism in a new way. At least as regards its claim and demand, the world that is accused of failure is becoming increasingly mundane and private, and simultaneously more universalistic and interventionist, and operates everywhere with a raised finger.

Human rights

According to a key thesis of the sociology of Émile Durkheim, the violation of a norm reaffirms and confirms its validity. The metamorphosis that is being fuelled by world risk society suggests how this insight can be updated and varied. Again, the protocol of worldwide failure – observations of extreme poverty and inequality, racism, the oppression of women, environmental destruction, the movements of refugees fleeing the new regions of 'barbaric' violence rooted in religious fundamentalism – transforms what was previously 'unthinkable' into the 'naturalness' of what is accepted as a matter of course. The litany of failure creates cosmopolitized forms of practice and spaces of action for political criticism and political activism. This is the language of the many cultural revolts (the Arab Spring, al-Qaeda, Occupy, or even the militant terror of IS), all of which have two things in common: they all came as a total surprise, and their aim is to change the world.

Anti-cosmopolitan movement

The often fierce resistance against the cosmopolitization of the world in particular – by the renationalization movements across the world, through the strengthening of the anti-European parties in France, Britain and Hungary, as well as in Germany – throws light on the power with which the world is becoming cosmopolitized. This cosmopolitization is in crucial respects hegemonic. It has intentional or unintentional carriers. Those underwrite the legitimacy of

democratic government, the cosmos of international organizations (the United Nations system, World Health Organization, World Trade Organization, International Monetary
Fund, etc.) and civil society movements and networks. But
there are also powerful sub-political carriers such as the
epistemic communities of experts, global businesses and
banking systems. They all increasingly penetrate localities.

The development towards a 'cosmopolitized modernity'
is always accompanied by its own problematization – that
is, by anti-modernity. Anti-modernity means *'constructed
and constructible certitude (hergestellte Fraglosigkeit)'* (Beck
1997: 63). Whether it be biologism, ethnic nationalism, neo-
racism or militant religious fundamentalism – it is always a
matter of ideologically dismissing the issues raised by the
process of modernization. Anti-modernity is in point of fact
a highly modern phenomenon. It is not the shadow of modernity but a fact coeval with that of industrial modernity itself
(ibid.: 36).

Thus, the cosmopolitanism of the world is determined
in a twofold way. On the one hand, it liberates all into the
freedom of new decision-making opportunities and constraints; on the other hand, the 'rigours of freedom' and the
hegemonic power with which cosmopolitization marches
onwards provide the starting point for anti-modern ideologies that insist on the alleged naturalness of nation, ethnicity,
culture, gender and religion.

The dialectic of cosmopolitization and anti-
cosmopolitization is played out in the political arena. This
also means that the failure to recognize the dangers of
climate change may not point to ignorance of this planetary
threat but, instead, to the fact that the recognition of the risk
to the global climate is driving the cosmopolitan metamorphosis of the world.

Risk society

[The] concept of *'risk society'* provides perhaps the most
striking example of recent theoretical attempts to make

sociological sense of such mapping projects and the reflexivity they exhibit. Put simply, *risk society* signals a new phase of modernity in which what were once pursued and fought over as the 'goods' of modern industrial societies, things like incomes, jobs, and social security, are today off-set by conflicts over what Beck calls the 'bads'. These include the very means by which many of the old goods were in fact attained. More pointedly, they involve the threatening and incalculable side effects and so-called 'externalities' produced by nuclear and chemical power, genetic research, the extraction of fossil fuels, and the overall obsession with ensuring sustained economic growth. *Beck* highlights the acute contradictions of a situation, where global risk and contingency follow directly from the modern drive to know, and through its knowledge, to control the world for human purposes. (Yates 2009: 20)

At the beginning of the twenty-first century, under conditions of *world at risk*,

the realm of ambivalence and uncertainty has struck back with a vengeance around an ever-expanding array of environmental concerns – climate change, declining fisheries, desertification, water scarcity, species extinction, and so on. All such issues demand active engagement and problem-solving on the part of experts and elected officials, just as they provoke criticism and dissent from activists. Crucially, science and science-inspired policy become principal targets of political controversy, with constituencies and special interests battling over the veracity of everything from global warming to peak oil to the risks of childhood immunizations to the dangers of genetically modified foods. (Ibid.: 20–21)

It is important not to confuse risk society with the catastrophe society. Such a society is dominated by the motto 'too late', by a fated doom, the panic of desperation. The small but important difference between risk and catastrophe – the anticipation of catastrophe to humanity (which is *not* catastrophe in reality!) – is a huge force of imagining, motivating and mobilizing. This way, again, risk society becomes a powerful actor of the metamorphosis of the world.

And we have to distinguish between 'global risks' and 'normal risks':

> 'Global risks' are 'different in kind' because they cannot be readily and 'naturally' understood as something 'unknown', in the sense of 'not yet known'. Rather, 'industrial, techno-economic decisions and considerations of utility' need to be understood as producing non-knowledge (*Nichtwissen*; see especially Beck, 2009; also in depth Wehling, 2006). Non-knowledge is not to be misconceptualized as knowledge that is (still) *absent*, something we do not know *yet*, in that it is not yet there. Rather, it needs to be understood as an unknown unknown, i.e. it captures the fact that there are things we do not know that we do not know. (Selchow 2014: 78)

To put it in other words: the notion of world risk society can be understood as the sum of the problems, for which there is no institutional answer.

Risk society is becoming the agent of the metamorphosis of the world. One cannot understand or deal with the world and one's own position in it without analysing risk society. Its conflict dynamic is a product of unprecedented dangers and unprecedented opportunities for political action. The one determines the other. Summarized as a model, which will be developed in the following chapters, this means the following.

A double process is unfolding. First, there is the process of modernization, which is about progress. It is targeted at innovation and the production and distribution of goods. Second, there is the process of the production and distribution of bads. Both processes unfold and push in opposite directions. Yet they are interlocked.

This interlinkage is produced not through the failure of the process of modernization or through crises but through its very success. The more successful it is, the more bads are produced. The more the production of bads is overlooked and dismissed as collateral damage of the process of modernization, the greater and more powerful the bads become.

It is only when the observer perspective brings both processes together that new possibilities of action open up. The focus on only one of these two interlocked processes makes it impossible to see the metamorphosis of the world. This is because the metamorphosis of the world is exactly the synthesis of these two processes and the realization of it through the observer. Hence, a theory and an analytical practice of metamorphosis together bring both processes centre stage and look at their interplay. The synthesis draws out a new diagnostical theory and concepts, such as 'global risk' (in opposition to 'normal risk'), 'cosmopolitization' (in opposition to 'cosmopolitanism'), 'risk-class' (in opposition to 'class'), 'emancipatory catastrophism' (in opposition to 'catastrophism'), 'relations of definition' (in opposition to 'relations of productivity'), etc. This enables us to observe the metamorphosis of the world. In fact, it enables us to understand the DNA of the world in that the interlocked double process can be imagined as a sociological equivalent to the double helix.

4   Cosmopolitan Theorizing

There is no shame in admitting that we social scientists are also at a loss for words in the face of the reality which is overrunning us. The language of sociological theories (as well as that of empirical research) allows us to address the recurring patterns of social change or the exceptional occurrence of crisis, but it does not allow us even to describe, let alone to understand, the social-historical metamorphosis that the world is undergoing at the beginning of the twenty-first century.

The word, concept or metaphor that I introduce in this book for this speechlessness as a distinguishing feature of the intellectual situation of the age is that of the metamorphosis of the world.

I use this theoretical diagnostic concept of transition to focus attention on the events that are unthinkable within the frame of reference of established social theories and on the new cosmopolitan framework and space of action. For example, climate risk, as well as other global risks, confronts us with the cosmopolitan *conditio inhumana*. But this also essentially includes the digital metamorphosis – that is, how life under the sway of digital totalitarianism is separated from life in political freedom by a rupture, by a global power of control that is transforming our whole existence (chapter 9).

Thus, the sociological theory of the metamorphosis of the world (as has been shown above) stands for the return of social history, for the message *history is back*! Therein resides – let us be frank – a provocation for mainstream sociology, and probably also for mainstream political science. The social theories of a Foucault, a Bourdieu or a Luhmann share one fundamental commonality with phenomenological and rational choice theories, notwithstanding all their mutual antagonisms: they focus on the reproduction of social and political systems and not on their transformation into something unknown and uncontrollable. They are end-of-history sociologies. They obscure the fact that the world is becoming metamorphosed into a *terra incognita*.

The theorizing of metamorphosis requires the metamorphosis of theorizing

Theories in the social sciences, in all their diversity, are in danger of losing sight of the historicity of modernity along with its sharply escalating destructive potential. Indeed, social history is truncated, on the one hand, into national history. On the other hand, the theoretical unpredictability and uncontrollability of the future, the dialectic of meaning and madness of modernity (Bauman 1989), is trivialized into the narrative of the rationalization and functional differentiation of the world. In this way, the horizon of sociology is underhandedly narrowed down and stipulated as being confined to the present. In other words, sociology falls into the trap of 'presentism', the stipulation and perpetuation of the present as without alternative. This ultimately leads to 'time-blind' and 'context-blind' models of modernization. The counterpart of this is the complacent belief that everything would be all right with the world if only people were like ourselves.

The interdisciplinary social and political theory of the metamorphosis of the world rejects this model of the reproduction of social and political order. This brings a whole series of new dynamics, processes and regimes of metamorphosis into view. For the sociological theory of metamorphosis, the central question is how the context of continuity and discontinuity, of the meaning and madness of modernity, can be conceptualized and studied empirically.

In later chapters of this book, I will offer a detailed treatment of what this means. Here I am concerned primarily with the technical matters of what this means for the work on theory: the theorization of metamorphosis calls for the metamorphosis of theory.

The customary understanding of theory in sociology, which equates theory with universalistic theory, distinguishes between theory and diagnosis of the present age. Implicit in this distinction is the value judgement that diagnosis of contemporary times is devoid of theory. As such, it is regarded as dubious. And, in fact, many diagnoses of the times overgeneralize isolated events or observations. But what I have presented and am presenting here is something entirely different. Similar to what I wrote in *Risk Society* and my book *The Cosmopolitan Vision* (or, in different ways, to the work of Anthony Giddens, Martin Albrow, Zygmunt Bauman, Bruno Latour, Arjun Appadurai and John Urry), what I am now saying is at stake is a theoretically informed, ambitious, historical diagnosis of the metamorphosis of the world. This develops a medium-range concept of process that allows us to describe the epochal change in horizon that universalistic theories fail to recognize.

This metamorphosis in the understanding of theory turns the hierarchical relation between universalistic theory and historical-theoretical diagnosis of the times on its head. The universalism in social theory characteristic of modern sociology that blinds it to the return of social history becomes a false universalism which – this is a key feature of my diagnosis of the times – obscures the cosmopolitized space and framework of action and the difference in worldview between thinking and acting.

But what requirements must a cosmopolitan turn in 'social theory' in the social sciences fulfil that shares neither the concept of 'society' nor that of 'theory'? Sociology is an observer but, at the same time, also a social 'actor' of the cosmopolitization of the world which it diagnoses. How, then, is 'theory' even possible? What does 'theory' mean?

Sociology is scientific observer and social actor of the cosmopolitization of the world

In the beginning is not the word, but the surprise. The surprise arises insofar as thinking and the articles of faith of the national worldview no longer insulate themselves against the experiences of the extended cosmopolitan framework and space of successful action. This occurs in everyday life, but also in business and politics, and not (or maybe indeed) in the final instance in the sciences.

Here I would like to show what is meant by the talk of cosmopolitized spaces of action and what role sociology plays in this, taking the example of how the transplantation of kidneys under medical supervision creates a kind of community of fate.

Our world is marked by radical social inequalities. At the bottom end of the global hierarchy are countless people who are trapped in a cycle of hunger, poverty and debt. Driven by sheer distress, many of them are willing to resort to desperate measures. They sell a kidney, a part of their liver, a lung, an eye or a testicle, thereby giving rise to a community of fate of a very specific kind. The World Health Organization estimates that 10,000 black market operations involving human organs take place each year (Campbell and Davison 2012). In this way the fate of inhabitants of the prosperous regions (patients waiting for organs) is connected with the fate of inhabitants of the poor regions (whose only capital is their bodies). For both groups, something that is literally *existential* is at stake – life and survival – in a very different meaning. The result is a modern form of 'dysbiosis': the amalgamation of two bodies spanning unequal worlds mediated by medical technology.

Continents, races, classes, nations and religions merge in the bodily landscapes of the individuals concerned. Muslim kidneys purify Christian blood. White racists breathe with the help of black lungs. The blonde manager sees the world with the eye of an African street child. A Catholic bishop survives thanks to the liver removed from a prostitute in a Brazilian favela. The bodies of the rich are being transformed into skilful patchwork assemblages, those of the poor into one-eyed or one-kidneyed storehouses of spare parts. The piecemeal sale of their organs is thus becoming the life insurance of the poor, in which they sacrifice part of their bodily existence in order to ensure their future survival. And the result of global transplantation medicine is the 'biopolitical world citizen' – a white, male body, fit or fat, in Hong Kong, London or Manhattan, fitted out with an Indian kidney or a Muslim eye.

We live not in the age of cosmopolitanism but in the age of cosmopolitization. This radically unequal cosmopolitization of bodies is not creating world citizens but is occurring wordlessly, without interaction between 'donor' and recipient. Kidney donors and kidney recipients are mediated by the world market, but they remain anonymous to each other. Their relationship is nevertheless an existential one, important for the life and survival of both parties, though in different ways. The simultaneous inclusion and exclusion of distant others – this is what I call 'cosmopolitization' – does not necessarily presuppose any dialogical connection or personal contact. Cosmopolitization, in short, can involve dialogue and direct interaction with 'others', but it can also take the form of a wordless, contact-free, asymmetrical relation (as in the case of kidney transplants or outsourcing capitalism inducing replacement between domestic and foreign labour).

These cases highlight the hallmarks of the *(in)human condition* at the beginning of the twenty-first century. No matter what people think – even if they define themselves as anti-cosmopolitans – if they want to be successful in their activities, they are forced to make sense and use of the cosmopolitized spaces of activities. This enforced and existential cosmopolitization is a *fact* and has to be clearly distinguished from normative cosmopolitanism. Indeed, it is a kind of imperialistic interconnectedness combining physically the radical unequal worlds.

'Fresh kidneys', organs transplanted from body to body, from the global South to the global North, are by no means the exception but are emblematic of a sweeping development. The social reality and notion of love, parenthood, family, household, occupation, employment, labour market, class, capital, nation, religion, state and sovereignty are in a process of cosmopolitan metamorphosis. The global others are here in our midst and we are simultaneously somewhere else.

On the conceptual level we have to distinguish between the observer and the actor perspective of cosmopolitization. The observer, the sociology of the metamorphosis, is making these invisible facts visible. Thereby, sociology is participating in processes of social construction. The role of public sociology might be (or become) to accompany the quantum leap from the national to the cosmopolitan outlook as a reflection process.

In order to determine the significance of cosmopolitan theorizing, it is useful to take up Robert K. Merton's (1968) distinction between 'grand theory' and 'middle-range theory'. It is not possible to conceptualize the metamorphosis of the world by following the universalistic understanding of theory, because the notion of universalistic theory excludes analytically what is at stake here – the change of universalistic assumptions. I suggest middle-range theorizing to conceptualize and analyse the metamorphosis of the world. As Anders Blok argues: we don't do theory, we do concepts. Middle-range theorizing meets and combines both empirical and theoretical ambitions in a workable cosmopolitan way.

> I think the argument can plausibly be made that a *cosmopolitan* social theory – as distinct from a social theory of cosmopolitanism – must *necessarily* be 'of the middle range' … Middle range, we might say, is not just an adjective, denoting a certain style of theorizing ('middle-range theorizing'), as Merton had it. Pushing further, it may also be taken as a noun (*the* middle range), connoting thereby an in-between epistemic meeting-place, an inter-crossing; or as a verb (*to* middle range), pointing to a dialogical process of mutual exchange across difference. In this extended sense, an aspiration to the middle range implies self-restrictions at the level of cosmopolitan theorizing: rather than search for a unified or universal theory, the challenge becomes one of forging a conceptual architecture capable of organizing ever-more meeting-points between diverse perspectives, as these grapple with collective and sharable experiences of encounter with 'the global other'. (Blok 2015: 112)

So there has to be an emplacement of theory cultures, of concepts that come from somewhere – Japan or Korea as a vibrant theory culture, or the US, or bringing out that this theory comes from Europe – this could be highlighted. Cosmopolitan theorizing needs to be imagined and organized as a dialogical 'space of emplacements', bringing different grounded histories back into social theory.

There are three dimensions of metamorphosis of the world.

- *Categorical metamorphosis* refers to the metamorphosis of seeing the world – that is, how global risks and cosmopolitan situations change the meanings of basic concepts of sociology – for example, from class to risk-class, risk-nation, risk-region; from nation to cosmopolitized nation; from catastrophe to emancipatory catastrophe; from rational capitalism to suicidal capitalism; from generations to global risk generations, etc. It is a process of metamorphosis of the world which is no longer embedded in paradigms of North and South, neoliberal notions of the 'West' and the 'Rest', but simultaneously includes the excluded global others in unknown trans-border relationships; they become objects of cosmopolitan theorizing (diagnostic, middle-range) and research.

- *Institutional metamorphosis* refers to the metamorphosis of being in the world. It is about the paradox of how working institutions fail: metamorphosis in the face of global risk produces a gulf between expectations and perceived problems, on the one hand, and existing institutions, on the other. Existing institutions might work perfectly within the old frame of reference. Yet within the new frame of reference they fail. Hence, a key characteristic of metamorphosis is that institutions simultaneously work and fail. This is what is meant by institutions hollowed-out through metamorphosis.
- *Normative-political metamorphosis* refers to the metamorphosis of imagining and doing politics, to the hidden emancipatory side effects of global risk. The main point is that talking about bads may also produce common goods, which means the factual production of normative horizons. But it does not come from universal values. It is grounded in empirical reality.

Part II

Themes

5

From Class to Risk-Class: Inequality in Times of Metamorphosis

Why metamorphosis of social inequality, why not transformation? Those who inquire into the transformation (or social change) of social inequality generally presuppose two things.

First, they conceive of social inequalities in terms of the distribution of goods (income, educational qualifications, social benefits, etc.); they do not even consider the distribution of bads (positions in the structure of the distribution of different types of risks), let alone the question of the relationship between the distribution, or the distribution logics, of goods and bads.

Second, their questions, thought and research move quite naturally within the framework of how goods are distributed in the national or international reference horizon. Thus, one sees the distribution of inequality through the spectacles of methodological nationalism which have become second nature. Another point is worth noting in this connection. The distribution of goods is in fact organized and observed at the national level. The distribution of bads – global risks – breaks with the national framework; bads become visible only within the cosmopolitan framework, in two respects: in order to explore theoretically and empirically the highly unequal forms of distribution and to search for political answers.

Both sets of background assumptions presuppose the reproduction of the national or international order of social inequality. Here the central question is: To what extent does social inequality increase or decrease in the national space, or internationally and globally? Since Marx, this has been translated into the conflict over the question of the extent to which the socio-economic classes and class antagonisms are abolished, preserved or aggravated. The French economist Thomas Piketty recently enjoyed worldwide success with his book *Capital in the Twenty-First Century* because, contrary to certain expectations, he attempts to prove that class relations were for the most part consistently reproduced through all the wars, and also the development of the welfare state, in the twentieth century.

Although both orientations – the focus on goods and the national-international duality – facilitate the study of the change in social inequality, they simultaneously fixate inequality research firmly on the pre-Copernican worldview in which the sun still rotates around the earth in matters of social inequality. By contrast, in what follows I ask how the Copernican turn in reflection and research on social inequalities can be accomplished.

Immured in the issue of transformation or change, the conventional sociology and economics of social inequality and class ignores the empirical reality at the beginning of the twenty-first century. It ignores the social and political explosiveness of global financial risks, climate risks, nuclear risks – in other words, the very metamorphosis of social inequality. By contrast, replacing the national with the cosmopolitan view brings the new realities – indeed, even more, the new dramas of the changing power relations and dynamics of social inequalities – into view: classes are metamorphosed into risk-classes, nations into risk-nations and regions into risk-regions.

1   Conventional Sociology Focuses on the Distribution of Goods without Bads

National class society is based on the distribution of goods (income, education, health, prosperity, welfare, large-scale national movements such as unions). World risk society is based on the distribution of bads (climate risk, financial risk, nuclear radiation), which are confined neither by territorial borders of a single society nor in time.

In order to clarify how I see the metamorphosis of social inequalities in the age of climate change, it is useful to draw a distinction to three other positions of conceptualizing social inequalities at the beginning of the twenty-first century. These positions can be distinguished according to the extent to which they accord central importance to (1) the reproduction or (2) the transformation of social classes with regard to (3) the distribution of goods without bads or (4) the distribution of goods and bads. The most interesting group here, as it is the most dominant one, is the group that concentrates on goods without bads and thereby focuses on the reproduction of class throughout the history of the twentieth and, maybe, the twenty-first century. As such it keeps practising the conventional sociology of class, ignoring empirical reality at the beginning of the twenty-first century – i.e., ignoring the social explosiveness of global financial risks, climate risks, nuclear risks: that is, the very metamorphosis of social inequality.

To be clear, the shift in perspective in class analysis that I am suggesting here is profound. This is clear when we recognize that the classics – Marx, Weber and Bourdieu – all focused on the production and distribution of goods without bads. They did not theorize risk as an explicit and systematic object of production and distribution. Given their historical context, this is of course obvious. Marx's focus is on the relation of exploitation. Weber concentrated on the relationship between power, market and change. Bourdieu was aware of the role played by economic and social risks in life; nevertheless, his analysis centred on different forms of capital, stressing the overall continuity of class relations over time.

In order to theorize and research the metamorphosis and radicalization of social inequalities in world risk society, I introduce the concept of *risk-class*. Risk-class sheds light on the intersection of risk positions and class positions.

Both the epistemological monopoly of class analysis in the diagnosis of social inequality and the methodological nationalism of the sociology of class have contributed essentially to the fact that established sociology is blind and disoriented in the face of the radicalized, transnational and risk-class society power shifts and equality conflicts we are witnessing today. The metamorphosis of class is happening. The theory and research of the metamorphosis of class in the social sciences has yet to begin.

The metamorphosis of theory and research starts by criticizing the nation-state bias. It is the scientific and public observation of and complaint about the lack of a cosmopolitan perspective that initiates the process through which a 'world perspective' (as it happens, especially in the small things and the pains of micro-political inequalities) becomes taken for granted. The new maps reporting violations against the demand for equality – no longer only at the national level but also worldwide – and the horrific images that lend this outcry visual form and disseminate it through the old and new communication channels, do the rest: the 'world', thought through to the distinction between global and local, becomes the horizon and at the same time the cosmopolitan object of study of social inequality.

When bads (risks and hazards), produced within the jurisdictional time-space of particular nation-states, transcend the boundaries of their legitimate authority, the second move or step of *metamorphosis* begins: a detailed record of failure. Bads and their impacts and costs are in effect constructed as non-existent by the construction of side effects; their impacts and costs are *externalized* to other populations, nations and/or coming generations and are thereby *nullified*. We suddenly discover that national borders are a key moment of metamorphosis because they have the power to determine which inequality is 'relevant' or not. On the one hand, they extinguish bads. On the other hand, seen as side effects, those bads grow with the speed and scope of modernization by cutting them off from institutionalized obligation, responsibility, law, politics, social science and public attention.

Their power of metamorphosis includes the 'politics of invisibility'. You do not 'see' the bads because you exclude the excluded. This way *metamorphosis* externalizes and neglects the bads. What does this mean? Two apparently contradictory moves of metamorphosis are taking place and can be observed: putting bads into both being (reality) and not-being (perception, acknowledgement) by concentrating exclusively on the production and distribution of goods.

As soon as we switch the 'world picture' and take the cosmopolitan frame of reference for granted we realize that the landscape of inequality has changed dramatically. It becomes apparent that the category of class, designed to incorporate the unequal distribution of *goods*, is *too soft* a category to capture the explosive realities of radicalizing inequalities in the context of globalizing expectations of equality and justice. This normative world horizon of global inequality implies an observer perspective, which includes the excluded victims beyond national boundaries. Then and only then the violence of climate change and its impacts, which might be called 'a continuous accident in slow motion', no longer slip under the radar of political and scientific attention.

The next move of *metamorphosis* is that, in the age of climate change, the notion of 'social class' becomes *Anthropocene class*. In other words, the issues and concerns about social inequality are becoming involved in the new geological era of earth's history.

The notion of the 'Anthropocene' and the notion of 'social class' and 'class society' belong to different worlds, maybe even to different epochs of history. We therefore have to raise the question: How and under which conditions does the inseparability of social class and the Anthropocene become possible, maybe even necessary? We have to look for empirical evidence of metamorphosis in the way people live and experience global risks (and how social researchers observe and describe this).

There is another set of questions to be asked: Where is the dominant defining power situated – in the logic of class (in the context of production and distribution of goods) subsuming the analysis of risk under that of goods? Why can risk not break through the logic of (national) class inequalities and conflicts?

But there is a complementary counter-question to be raised as well; here the defining power lies in the logic of risk subsuming the analysis of class inequalities and asking: How are global risks changing the logic of national class inequalities?

The first question is part of the narrative of *continuity* stating that industrial success has been forced to reveal its catastrophic underside. The counter-question is about the narrative of *discontinuity* and *metamorphosis* stating that the anticipation of climate catastrophe to humankind is creating ('post-class') inequalities and conflicts of the worst kind. In both cases the notion of 'risk-class' is central, but in the second case the dominance lies not in risk-CLASS but in RISK-class.

The next set of questions perceives and analyses these patterns in the light of what is expected to be a just and fair distribution of goods *and* bads in the age of climate change. If the first set of questions is centred on issues of distribution of risks, this second set broadens the perspective by focusing also on matters and patterns of process and production, of institutional arrangements and existing laws, perspectives of politics, and social scientific knowledge production. Metamorphosis here means shifting the perspective from descriptive patterns of inequality, which are treated as simply 'given' and therefore as 'problems to be managed', to concerns of injustice.

In this way the (factual) normativity of the notion of 'world' comes in. This frame of and focus on (in)justice can help us understand why patterns of inequality exist and are being interconnected with all kinds of social, economic and political conditions inside and outside the national container. Here, then, the metamorphosis of theory changes the issue from describing to explaining inequalities and injustices.

This is the point of metamorphosis: the combination of risk and class is not obviously apparent 'in' looking at a natural disaster. It is only apparent when the normative horizon of 'social justice' – i.e., critique! – comes into play. So, it is again only the applied normative horizon of 'social injustice' that makes 'risk-class' visible. A good example is the case of Hurricane Katrina in 2005. As documented in the literature (e.g., Walker and Burningham 2011: 217), it took the devastating and shocking racist impacts of Katrina actively to shift the perspective from the natural occurrence of the flood and its material destruction to the issue of inequality of risk-classes.

In other words, there are two implications in this: it was *not* the descriptive perspective on the disastrous social implications of flooding that awakened the attention to normative issues of justice, but exactly *the other way around* – only the lived experience of the 'racist flood' opened eyes and minds to the issues of injustice *and* unequal distribution of floods and flood risks.

Social inequality and climate change combine a number of different dimensions. Climate change as a physical process must be understood as a power to redistribute radical social inequalities. It alters the timing and intensity of our rains and winds, the humidity in our soils and the sea level around us. Because of this redistributional power, climate change is a social and natural challenge as well as bringing up the issue of justice. It is about who gains and who loses as change occurs and its interventions to moderate change unfold. Therefore it is not only climate change as a physical process but also political climate responses and the discourses around them which introduce – produce and reproduce – old and new social inequalities. Of course, the issue of 'vulnerability' has become very common. But not much notice is taken of those important new measurements of social inequality in the sociology of class. Here a specific aspect of metamorphosis comes into view: the ecological conditions, distribution of assets and systems of power that place certain populations or communities, or even continents, at greater risk in the phase of changing climate. There is a basic condition of politics implied: Who is put at risk in the phase of policy responses? So there are two resources for how climate changes inequalities: material injury and violations following the changing climate weather patterns and inequalities as an outcome of climate change interventions. Often injury is hereby added to injustice: those radical inequalities are not recognized on account of the 'politics of invisibility' (chapter 5).

2   Coastal Flood Risk and *River* Flood Risk

The shift of 'at-risk-spaces' is a good case that illustrates the above points. In order to analyse the uneven patterns of flood risk you have to make a decision about the unit of research: What are the *at-risk spaces*? Those at-risk-spaces can be defined either socially or geographically. Looking only at geographical patterns of flood risk, there are two units of research possible – (1) coastal flood risk or (2) *river* flood risk – which might be interconnected physically but don't have to follow the same pattern of social class inequality.

Several studies of patterns in who is at risk from coastal and river flooding in England, Wales and Scotland have recently been completed (Fielding and Burningham, 2005; Walker et al., 2003, 2006; Werrity et al., 2007). Each of these takes a similar form, focusing on identifying patterns of distributional inequality. Geographical Information System (GIS) and statistical methods are used to relate the spaces indicated on official Environment Agency maps as being at flood risk (from river and coastal sources, not pluvial) with social data from the census. The concern has predominantly been with patterns of social class and deprivation. Different methods for how populations are allocated as being inside or outside of the floodplain are explored by Fielding and Burningham (2005), demonstrating some of the dependencies of outcomes on the methodological choices that are made, and the uncertainties that exist within such analyses.

Walker et al. (2006) have undertaken the most developed and involved analysis and this contains some striking evidence. In England there are 3.3 million people that live within the Environment Agency specified zone delineating a

1% or greater annual probability of flooding from rivers or 0.5% or greater from the sea. Divide this population across ten deprivation categories (for deciles) from the 10% most deprived areas in England to the 10% least deprived, and a profile of flood risk against level of social deprivation is produced. (Walker and Burningham 2011: 219–20)

These empirical studies seem to affirm the general claim that the production and distribution of risk does *not* transform but reinforces the logic of class distribution. There is evidence for this in a huge amount of literature beyond the flooding in England. Researchers who study social vulnerability see economic class as a central aspect of vulnerability to loss in a catastrophic event (Cutter et al. 2003; Cutter and Emrich 2006; Oliver-Smith 1996; Phillips et al. 2010). A lack of financial resources has a direct impact on one's ability to maintain both residence and lifestyle that reduces vulnerability and to prepare when the threat of disaster is eminent. One of the more comprehensive and macro-level analyses of social vulnerability (Cutter et al. 2003) found that eleven factors explain 76 per cent of the higher vulnerability of some US counties compared with others, including a lack of personal wealth, single-sector economic dependence, housing stock and tenancy (proportion of mobile homes, renters, and urban location), proportion employed in lower service occupations, and infrastructure dependence. In this literature it seems to be universally evident that economic disadvantage on both the individual and the community level leaves some populations more vulnerable to the most severe impacts of flood disasters, etc. But this again doesn't hold if you distinguish between different geographies of flooding.

You might discover an important difference when the data on flooding are disintegrated into separate units – river flooding and sea flooding zones. Then it becomes what the 'Anthropocene inequality' implies: the variation of the geographical unit of research shifts the perspective.

When we look at *coastal* flooding, class differences are evident. If we look at *river* flooding, the class differences nearly disappear.

Looking at the profile for...river flooding this is very flat, with little variation from the most to the least deprived. And there is an important political difference in that: flooding which only or mainly hits the unprivileged may be an unpolitical flooding framed and lost in the darkness of side effects. But river flooding which hits privileged parts of the population is (in England) a highly political flooding of Tory-voters' home counties. This then breaks down the 'Berlin Wall of side effects' creating close to election times a central issue for a Tory government. This demonstrates that with risk class there is a new kind of enmeshment between the flood and the state.

We need to recognize that thinking about the uneven impacts of flooding in terms of distinct population groups is problematic. Some of these categories intersect in complex ways (for instance disabled people are disproportionately likely to be poor, as are members of minority ethnic groups, women and older people); not all within them are equally vulnerable and vulnerability is a dynamic rather than a static quality (people can move in and out of vulnerability). Thus while the talk of 'vulnerable groups' often provides a useful shorthand to focus on the uneven impacts of floods, this framing needs to be used with some caution. (Walker and Burningham 2011: 222–3)

So far we have seen an interpretation of 'risk-class' where class positions and risk positions (more or less) correlate. The *metamorphosis* to 'natural class' gets started but stops with the logic of class conflicts dominating the logic of risk. But this is not the whole story. In world risk society the logic of global risk metamorphoses the logic of class. Let me give you two examples for this.

3   Climate Risk is Convulsing the 2,000-year-old Civilization of Winegrowing in Southern Europe*

The first example demonstrates that the distribution of risk does not follow the logic of class but that the opposite is the

---

* On what follows, see Fichtner (2014).

case, and it exemplifies the notion of 'Anthropocene class' as well. Someone who asks 'Can I see, hear, taste, smell, touch climate change?' will receive in response a resounding 'yes and no'. On the one hand, extreme weather events are on the increase. A winegrower who made it to number 273 on the list of the richest French people and regards climate change as a *faux problème*, a phoney, manufactured problem, recalls a storm of devastating strength. Within the space of a couple of minutes, 5,000 hectares of the best Bordeaux vines were 'literally hacked to pieces', says the wine-maker, for whom the vineyards are the source of his wealth and his social prestige. 'Perhaps, monsieur, this is your climate change' (Fichtner 2014).

Here a risk-class position is developing in which climate risk affects the wealthy – in this case, very unusual wealthy people, the owners of the vineyards which produce the best wines in the world. Global climate risk is transforming the class hierarchy, therefore, turning it on its head and simultaneously binding it into society's relationship to nature (vineyards).

However, climate risk should not be confused with catastrophic climate change. That would be a serious error in several respects. On the one hand, winegrowers (like all losers of climate risk) are dealing with catastrophic natural events whose man-made character is not plain to see. As I said before, as a general rule, in particular cases it is impossible to make a clear distinction between 'natural' and 'man-made' natural disasters (and even the scale of the natural disaster is ultimately irrelevant in this regard). This interpretation that the disasters are man-made, and hence can be attributed to decisions, production processes, road traffic, air traffic, political inaction, etc., arises only in memory, in statistics, in public discussions about climate change. A wealthy winegrower who accepts this interpretation is living in a different world. Two thousand years of civilization is under threat. However – this is a second important orientation – the winegrowers are dealing not with an actual climate crisis but with the anticipation of such a crisis, as a looming future event that poses an insidious threat to humanity – hence, with climate risk.

> Extreme weather is becoming more common in all of France's winegrowing regions. Heavy rains and hailstorms frequently come on the heels of summer heat waves and dry periods. Winters and nighttime temperatures are so mild that the plants are never able to rest. Few winegrowers continue to deny these tangible phenomena.
>
> On the other hand, it isn't easy to perceive that the last three decades have been the warmest in the last 1,400 years. It's hard to comprehend that the average annual temperature has increased by 1 degree Celsius (1.8 degrees Fahrenheit), that the Atlantic and the Mediterranean are warming almost imperceptibly, and that days are getting just slightly warmer. Human beings lack the natural sensors to detect such changes, but grapevines have them. The vines are suffering from ongoing stress, say some vintners. Vineyards are in turmoil, not just in France but also in Italy, Spain and all of Southern Europe – in all the places where it has always been warm and where it is now getting too hot.
>
> ...
>
> Throughout southern France, vintners' calendars are in turmoil. The ripening periods are constantly getting shorter and the harvest season is arriving ever earlier. Grapes were picked in October in the 1960s...but now the harvest is moving closer to early September. The tried-and-true experiences of old vintners have become meaningless, and traditional rules of thumb, based on decades of weather observations, are now invalid...'You can regret everything,' says Guigal [one wine grower], 'or you can roll up your sleeves and get to work.' His plan is to reinvent French wine. (Fichtner 2014)

'Climate change' from the upper-class perspective of producers of world-class wines therefore does not appear as climate change but presents a situation that requires a decision. Those who deny the existence of climate change (perhaps in the belief that they are thereby securing their wealth and conserving tradition) run the risk of accelerating the destruction and of failing to take advantage of necessary alternatives. Those who acknowledge the existence of climate risk in virtue of this alone devalue their possessions, on which the shadow of rapid future destruction now falls. For a start, this has tangible (also invisible) economic consequences. The social perception that the precious vineyards in the south of Europe are endangered undermines their economic value, even though catastrophic climate change represents only a future threat. At the same time, however, the recognition of climate risk first makes it possible to take countervailing measures within one's own scope for action. However, these counter-measures are taken in the context of a different worldview – namely, within the cosmopolitized framework of action, which includes society's relationship with nature as well as the global climate models and the failure of politics.

The multifaceted concept *terroir* plays an important role in this worldview. What might be carelessly translated as 'earth' or 'ground' is actually at once an admixture of new and old natural objects and cultural processes, history, law, belonging, demarcation, identity and even 'global climate' (as in invented notion of climate scientists). *Terroir* articulates the traditions, changes and threats that are giving rise to a short circuit between the fate of the world and the fate of one's own life. This is at the root of the fears of the winegrowers, because *terroir* stands for property, status and the typical character of their wines, hence for branded goods that open up world markets, thus ensuring their wealth and status.

The natural class position of the wealthy Southern European winegrowers is being transformed within the framework of climate risk into a twofold, cosmopolitan specific field of action. On the one hand, climate risk is bringing about a passive, coercive, painful cosmopolitization, which is fundamentally transforming the winegrowers' conditions of work and action. The weather, which has mutated into the agent of climate change, is becoming an everyday enemy that is threatening to transform the sources of their wealth and identity into deserts. Here, 'the weather' does not stand on its own, but behind it are the actors who produce and deny climate change. They represent an existential threat which does not obey the old friend–enemy logic but, instead, bridges national, religious and ethnic divisions; nevertheless, they appear as a bundled power that plunders the foundations of our natural, historical, moral and economic existence.*

On the other hand, this also creates practical possibilities for reinventing French wine:

> The strategies of working in a vineyard have changed radically, and vintners know more than ever about ripening processes. In fact, the vintner's toolbox is well equipped to react to changes in the climate. The biggest challenge is certainly replacing grape varieties....
>
> But is it getting too warm for Grenache in Châteauneuf? Why not plant Syrah, the variety popularized around the world as Shiraz? What's wrong with growing late-ripening Cabernet Sauvignon farther to the north in the Rhône Valley? Or perhaps even in Burgundy? Why not move the vines farther uphill to cooler altitudes? Or plant them on the north faces of hills to avoid the sun? (Fichtner 2014)

Special grape varietals that cohere with the philosophy of a biodynamic worldview can open up new world markets. Thus, Isabelle Frère, a winegrower 'who only wanted to live a sustainable life, has completed a minor, ironic miracle of globalization, because she now sells most of her wine to Japan' (ibid.). Those who remain wedded to the national view are the losers; by contrast, those who take the leap into the cosmopolitan space of action can create opportunities for securing local traditions and livelihoods. But the term 'climate change' contains a special kind of planetary threat: 'names that once represented individual worlds, worlds almost painted in oils: the Loire and the Rhône, Burgundy,

---

* It may be important in this sense to note that Bruno Latour, whose work is marked by an enormous intellectual and political commitment to combating climate change, is the scion of one of the leading French winemaking families.

Bordelais and Champagne. Appellations like the lines of a poem: Médoc, Pomerol, Pauillac, Meursault, Chablis, Hermitage, Pommard. Forgotten. Consumed. Finished' (ibid.).

4   How Privileged Places turn into Risk-Places

The second example is the case of the metamorphosis of goods into risks. Significant maritime and industrial areas are characterized by New York City as places along the waterfront that contain dense clusters of industrial firms and water-dependent businesses. In 1992 they were designated as privileged areas to be protected and encouraged for continued use in this fashion. These privileged positions changed into risk-positions when viewed and mapped through the eyes of climate justice experts: in this horizon of anticipated catastrophe, the producers of goods became *victims of bads*, potentially or in reality because those experts realized immediately that every single one of these privileged areas was located in the flooding and storm zones.

Those privileged areas and potential victims of floods and storms become objects of *metamorphosis*, their mode of existence changed from producers of goods into producers of bads threatening the nearby communities.

How did this happen? Looking at those industries and businesses through the eyes of climate risk inequalities and injustice, there is a different set of questions to be asked and answered shedding light on different realities: the logic of the production and distribution of bads now dominates the production and distribution of goods. Thus those climate experts cited a number of chemicals present in these areas, chemicals to which the community would be exposed during flood and storm events. They include, for example, trichloroethylene, which is a carcinogen; naphthalene, which causes liver and kidney damage and harm; and n-hexane, which affects the brain, etc. This is not about a different cultural perspective; this is real, on account of the reality which is uncovered in the light of anticipated climate catastrophes. Since past experiences with disasters such as Hurricane

Sandy destroyed both – the irrelevance of side effects and the security measures to contain chemicals – the goods-producing industries change into producers of bads to vulnerable communities. And in this horizon of expectation the government-organized irresponsibility collapses. Knowing that these floods and storms are going to become more frequent and more intense, and knowing that there are vulnerable communities, to do nothing, not to act, is an abdication of democratic government legitimation. To become politically active then becomes an issue of power and legitimation.

Climate risks shift the notion of risk-CLASS into RISK-class. Here the essential risk to humankind comes into focus (as is also the case with nuclear meltdowns, which are of such a scope that rich and poor, North and South, are threatened alike).

The shift from risk-CLASS to RISK-class depends very much on the profile of the words 'future' and 'justice'. As I said before, you can look at present flooding and class from the past. Thereby you unconsciously take the frame of reference of nation-state and class society for granted, and then you see flooding dominated by class (risk-CLASS).

In order to understand how the production and distribution of risk metamorphosed class inequality, the euro crisis and its consequences is a good example. What happened was a transnational inequality dynamics that created the division between creditor and debtor nations, between the European North and the European South. This led to a hierarchy of countries, a hierarchy of 'risk-nations'. The group of Southern countries descended to a second 'risk-class' position, while Germany became an 'accidental' empire. At the same time, there was a dramatic overall descent of certain social groups within the debtor countries. Pensioners, the middle class and the young generation experienced a dramatic worsening of their economic situation. This dynamic of risk production and distribution binds together the structure of inequality on the European, transnational and international level with the inequality dynamics within nations. This cannot be grasped with the category of 'nation' as we know it but can be captured with the concept of 'risk-nation', just as the consequences of climate risk cannot be captured any longer with conventional concepts such as 'regions' but needs to be approached with the concept of 'risk-regions'.

5 Outlook

Taking the above, we see three main points.

First, the cosmopolitan perspective shifts the focus away from seeing people and communities only as potential vulnerable *victims* towards seeing them as *citizens* with rights to be asserted, achieved and protected. If climate disasters are perceived as a matter of justice, then it is necessary to ask whether the patterns of inequality and vulnerability that exist are just – rather than treating them as risks to be managed.

Second, the cosmopolitan perspective raises the '*who*' question, because the unit of research and political action is no longer predefined and 'given' by geographically defined risk areas (of flooding, etc.) in presupposed nation-state borders. The 'who' question transcends these 'given' walls and borders of thinking and seeing. By concentrating on the production and distribution of bads, you have to include their point of impact, which is obviously not tied to their point of origin, and you have to look at their transmission and movements, which are often invisible and untraceable to everyday perception. In order to overcome this social invisibility (constructed as side effect), the unit of research must connect what is nationally and geographically disconnected. This is exactly what the cosmopolitan perspective aims to do ('methodological cosmopolitanism').

Third, besides the 'who' question, the cosmopolitan perspective can help to understand the '*why*' question – why patterns of climate inequality exist – by making connections between the two different forms of risk-CLASS and RISK-class. This way the nature of global risk in opposition to the nature of goods comes into focus. Goods are things –

machines, buildings, bodies, foods, educational degrees (PhD, etc.). Global risks are of a completely different nature; they are socially constructed in knowledge – anticipations, imaginations, probabilities, possibilities, aspirations corresponding to different kinds of imagined apocalyptic catastrophes. Thus the politics of global risk is, first of all, intrinsically a *politics of knowledge*, which brings up questions such as (1) Who is to determine the harmfulness of products and technologies of the involved risk and their dimensions? Is the responsibility on those who generate those risks or those who benefit from them? Are those affected or potentially affected by them included or excluded? (2) What is to count as a sufficient proof and – in a world in which we necessarily deal with contested knowledge or knowledge that we don't know and will never have in the classical sense – who decides this? (3) If there are dangers and damages, who is to decide on compensation for the afflicted and who takes care to ensure that future generations are confronted with less existential risks?

6

Where Does the Power Go? Politics of Invisibility

This chapter addresses the problematic of the metamorphosis of power in world risk society. Under which conditions do normal risks turn into global risks and vice versa? How does the architecture of power relations in world risk society metamorphose? And who holds the resources of domination that enable social definitions and redefinitions to be established?

Using the theory of metamorphosis as a prism to analyse the historical transformations in key social and material relations, fixing on existing power relations, it is necessary to introduce a new time-diagnostical concept to examine the categorical and institutional metamorphosis of power – the *relations of definition as relations of domination*. This middle-range concept, which could become the focal point of cosmopolitan theorizing and research, cuts through the superficial 'rationality' of risk assessment and management and opens the outlook on the underlying power structures and agencies of the social definition of global risk nationally and globally.

In this way, the perspective of metamorphosis shifts the focus on power and domination from 'power relations of production' (in the Marxist sense) in modern global capitalism to 'power relations of definition' in world risk society.

With 'relations of definition' I refer to the resources and power of agents (experts, states, industries, national and international organizations), the standards, rules and capacities that determine the social construction and assessment of what is a global risk and what is not. Among these are the politics of invisibility, the standards of proof, and standards of compensation. To what extent can imperceptible risks (such as nuclear radiation and climate change) be made publicly invisible and unobservable? To what extent does the politics of invisibility produce a situation of not knowing the existential risk?

There is a second aspect of the metamorphosis of power: Who defines risks as 'global' or 'normal'? And which symbolic strategies and definitional means are applied?

Institutional metamorphosis refers to the metamorphosis of being in the world. This is exemplified here by analysing the paradox of why and how functioning institutions fail. Only if the lens of social change is replaced with the lens of metamorphosis does the view open to the new cosmopolitized field of activities. This metamorphosis of power – from conceptualizing and discovering the conditions of definition interrelated with and simultaneously decoupled from the relations of production – becomes the centre of cosmopolitan theorizing and the unit of empirical research for methodological cosmopolitanism. It is a 'positive problemshift' (Lakatos 1978) because it throws light onto the institutional metamorphosis by examining why, faced with manufactured not knowing about existential risks to humanity, the nationally bounded and grounded juridical or legal standards and the universal scientific norms of causality simultaneously function and fail.

The perspective of social change thereby misses the historical metamorphosis of power, including the shifting relationships between national law, principles of justice, equality, parliament, government and experts ('epistemic cultures').

There is also a normative-political metamorphosis that has to be taken into account: the imperative of democracy and justice applied to the power relations of definition makes visible a metamorphosis of *revolution*: the 'revolution' centred on the power relations of definition does not happen where the Marxist notion of revolution expected it to happen. There is, for example, no need to support revolutionary movements, as the left used to do in Latin America, but to collect money, for example, to distribute dosimeters to the poorest and most vulnerable parts of the risk population. Thus the monopoly of power incorporated in the relations of definition can be changed by rather small interventions. It is a matter not of overthrowing the power relations of production (socialistic revolution) but, to start with, of giving all individuals the dosimeter – i.e., the means to determine the doses of radiation incorporated in their living conditions.

1 Politics of Invisibility

Global risks are fundamentally characterized by the problematic of invisibility. This problematic of invisibility is intrinsically connected to the problematic of power. In order to analyse the new landscapes of relations of definition, it is useful to introduce a time-diagnostic dualism between a *natural* ('given') invisibility of highly civilizational risks and *manufactured* invisibility (politics of invisibility).

The quintessential risks of world risk society – for example, climate change, risks associated with nuclear power and financial speculation, genetically modified organisms, nanotechnology and reproductive medicine – are increasingly complex in their courses and effects (replete with synergetic and threshold effects) and temporarily and spatially expansive in their scope. Because of their complexity and the time lag they are – paradoxically – characterized by *natural* invisibility: paradoxically, the more complex the production and nature of risks become, and the more production and definition of risks depend on global interconnectedness, the more 'natural' the invisibility of those risks is.

It was the 'anthropological shock' in the aftermath of the Chernobyl catastrophe as a media event which made the invisibility of the radiation risk visible (Beck 1987). Based on the direction from which the winds blew the 'radiation cloud' westwards, whole populations in Europe – beyond class and nation – experienced that, in the existential questions of their own life and the life of their children, they depended totally on media representations, narratives, experts and anti-experts quarrelling with each other: they depended as well on technological equipment, maps, rumours and competing theories introducing a vocabulary to everyday life they didn't understand. Not all risks – not local ones (such as a smoking chimney) – are characterized by a state of natural invisibility; it is those that are globally produced, distributed and defined. Without the information provided by the media and other social institutions, citizens are not even aware of the risk to their life and to the lives of their children and neighbours. There is no direct component to one's experience of global risk, no simple sensory and common-sense evidence. Global risks (such as radiation but also climate change) not recognized scientifically do not exist legally, medically, technologically or socially, and they are thus not prevented, treated or compensated for.

Natural invisibility implies and multiplies the institutional power of risk definition. As long as the citizen does not possess the means to make the invisible threat to their life visible, all the power to define global risks lies in the 'hands' of the institutions (experts and legal systems, industries, government, etc.). As we will see later, in the context of global and national risk definition the afflicted risk nation-state can play an important role.

There is a remarkable feature of global risks: they introduce the twofold existential threat, first, to the life and sovereignty of the citizens and, second, to the authority and sovereignty of the nation-state. Not only the state, but even the possibility of the state, depends fundamentally on guaranteeing the safety and security of its people. A government that admits and acknowledges its failing in the face of global risks either threatens its legitimation and existence or gets involved in a metamorphosis of politics (an example of the latter is the turn in nuclear energy politics in Germany after the Fukushima disaster). This implies that politics of invisibility is an important strategy to stabilize state authority and the reproduction of social and political order by denying the existence of global risks and their effects of 'ecological and risk appropriation' and health effects to large parts of the population.

In the process of manufacturing invisibility – i.e., in the politics of invisibility – natural invisibility can be instrumentalized. Actively doing nothing is the cheapest, most effective, and most powerful political strategy to 'simulate' the controllability of uncontrollable risks and open-ended catastrophes such as radiation and climate change.

The nearly complete public disappearance of invisible risks is not unique to specific political systems, for example, the Soviet Union after the Chernobyl disaster. You find those practices in Western democracies as well. Here, too, institutions designed to control risks simultaneously fail and don't fail. They fail because they have no idea or answer as to how to cope with these global risks. They do not fail because their politics of invisibility is continually making exactly those risks invisible to the public. Here we can observe something which could be called the 'functionality of failing' or the 'functionality of dysfunctionality'.

In all countries of the world all kinds of imperceptible risks 'are continually being made invisible by the industries that produce them, and that in turn are aided by administrative bodies that do not regulate them. The tobacco industry has infamously worked to make the health effects of smoking publicly invisible' (Kuchinskaya 2014: 159–60). But this is at the same time a historical example for the politics of metamorphosis: throughout long-lasting national and global battles, the power and politics of invisibility has been overcome and turned into a politics of visibility, demonstrating that even most powerful industries can surrender and be forced to acknowledge the health risk of smoking to large parts of the population.

The chemical industry has been waging a campaign against recognizing the health and environmental effects of pesticides... Historical and sociological studies have documented the various strategies used by industries to displace dangerous toxins as objects of public attention: reframing the public debate on hazards, promoting fake debate where there is a scientific consensus, silencing critics, orchestrating studies to counter even strong evidence of harm, blaming victims' genetic makeup or lifestyles and denying environmental influences and presenting a lag of monitoring as an absence of health effects. These strategies appear in cases of hazards created as a result of accidents as well as in cases of routine production of hazards. Indeed, even climate change is a complex phenomenon that cannot be perceived directly, that needs to be made publicly visible, and that some interests are trying to make invisible. (Ibid: 160)

But we also have to draw limits to the perspective of metamorphosis: there is not one truth upon which the broadened analysis of the power landscapes of risk definition can rely. It is not the God's eye truth that is the point of reference for the analysis of metamorphosis. But it is the shift of perspectives by which the risks are rendered more publicly visible and observable. The focus of analysis shifts to the dynamic of the public recognition of imperceptible risks depending on the power relations of definition. Public visibility depends on whose voices can be heard and which groups 'possess' the means of institutional and infrastructural support (research).

In order to broaden the range of strategies of invisibility, it is useful to draw a distinction between *catastrophe*, unlimited in time, space and people affected, and the notion of *accident*, defined by strict limitations in time, space and people affected. Confusing open-ended catastrophes with limited accidents means obscuring imperceptible with normal risks (nuclear radiation with car accidents). From this arise strategies of symbolic politics: unlimited catastrophes are being framed and 'managed' as limited accidents.

On the one hand, risk research is initiated but, on the other, organized not to ask the vital questions; the scope and the research in time, space and affected social groups are limited. Global risks or nuclear risks are defined by numbers of death causalities, excluding all those who are still alive but suffer severe health problems; excluding effects on non-living future generations; applying orthodox norms of causality; limiting the 'zone of alienation', where there have to be administrative efforts for 'rehabilitation', including state radiological control of food, health and recuperation programmes for children; regulating the observation of the changing life conditions of risk positions and populations; and pushing for selective approach to radiological data. From this they limit the scope, efforts, costs and specialized treatment for programmes, thus creating both our not knowing and the expansion of the ongoing catastrophes – of course, not forgetting the sophisticated strategy of not looking for pragmatic answers.

There is another very effective strategy of invisibility: by shifting the focus (frame of reference) from health effects to economic costs and economic-administrative problems – that is, by emphasizing economic constraints.

> This might appear as a paradox: One consequence of the Chernobyl accident…was massive radiological contamination, yet the topic of radiological contamination has never been the main focus of the related media coverage. Instead, more than 90 percent of all the articles…discussed socio-economic issues.…discussions of how to live in the contaminated areas and who should be evacuated were directly tied to questions of funding and the hope for international assistance. (Kuchinskaya 2014: 91)

2 Manufactured not Knowing

Our knowledge of global risks is highly dependent on science and experts. They are the central power institutions in a world where everyone is confronted with invisible existential risks beyond control. But science and experts in risk society play an increasingly paradoxical role which seriously undermines their legitimation and power. For example, nuclear industries and experts occupy a Janus-like position: they are at the same time both creator and assessor of risk. This undermines their power position based in the relations of risk definition.

This is especially true in the case of nuclear industries and experts. Hans Blix, director-general of the IAEA from 1981 to 1997, argued five years after Chernobyl: 'The future of nuclear power depends essentially on two factors: how well and how safely it actually performs and how well and how safely it is perceived to perform' (Fischer 1997: 171). Thus the epistemological community of internationally organized nuclear safety experts tried to affirm their control over the areas of uncertainty by re-establishing the boundary between experts and the public, especially the affected population: rational, scientific conclusions of the experts juxtaposed with the 'irrational, uneducated, emotional, and some time even hysterical' public.

Metamorphosis is deeply connected with the idea of un-awareness, which embeds a deep and enduring paradox. On the one hand, it emphasizes the inherent limitations to knowledge, especially the reality that some knowledge is knowable or does not attract a willingness to know, that nano-technology, bio-engineering, and other types of emergent technology contain not only knowable risks but also risks we cannot yet know, providing a window of fundamental limitations to society's ability to perceive and govern risks. This state of 'reflexive' unawareness poses key challenges not only to risk research (such as weather) to proceed with the potential of genetic engineering, reproductive medicine and their applications. It is much more than that. It is the coincidence, the coexistence of not knowing and global risks which characterizes the existential moments of decision not only in politics and science but also in everyday life situations. How to survive and decide under conditions of not knowing and unawareness is not a fictitious but the real existential problematic at the beginning of the twenty-first century throughout all levels of decision-making, in families and in national and international organizations.

How do I deal with the unimaginability and imperceptibility of life-threatening risks in everyday life? There is a specific kind of individualization process. 'More than two decades after the accident, the paradoxical fact about Chernobyl radiation is that individuals are responsible for their own internal contamination doses...In other words, people make their own doses, but not in circumstances of their own choosing; these circumstances present a unique intertwining of radiological, geographical, economic, cultural, infrastructural, and other factors' (Kuchinskaya 2014: 39).

Asking affected lay people to make individualized style changes does not address the underlying structural impossibility of this 'risk position'. But radiological contamination is only one part of the known condition of unawareness and danger. The other part is that individuals and families have to develop practices of how to cope with the invisible danger depending on conflicting experts' information in order to mitigate the radiation risk and reduce their own doses. This is 'hard work' and has very different outcomes depending on the socio-economic situation and the local infrastructural relations of definition. Some people say: 'I would rather die from radiation than from hunger' (Kuchinskaya 2014: 40; cf. UNDP et al. 2004: 33). As we can see from this statement, the range of options depends very much on the economic situation. Here again we find the inequality of 'risk-CLASS'. Or, to put it in more theoretical terms: in this case the relations of definition are subordinated to the relations of production.

To acknowledge not knowing is one thing, but to live with invisible unknown risks is another. As Olga Kuchinskaya's research shows, it is wrong to expect that the affected groups are sharing the same perspective on radiation danger and are practising the same answers. It is also wrong to present their views as irrational, intuitive and experiential, and as not changing over time. The opposite is the case: 'One individual can hold multiple perspectives' on radiation risk.

Most individuals can express different perspectives, and they can – and do – change what they argue depending on the context and with whom they are talking. For example, the same person might argue that Chernobyl has had 'grave health consequences' in some contexts (e.g., claiming Chernobyl benefits, teaching children, or talking to local administrators) while assuming a position of indifference in his or her daily life. (Kuchinskaya 2014: 41–2)

And the risk positions people are in are 'Anthropocene positions' as well.

Difficult economic conditions lead to a greater dependence on free resources, including private garden plots, forests, and wild pastures. At the same time, post-Chernobyl radiological contamination tends to accumulate in the top layer of a forest or field, and a number of practices effectively transfer radionuclides into private plots. For example, residents send their cattle to graze in wild pastures, then they consume radionuclides with milk and other dairy products – unless they have access to cultivated pastures purposefully planted with special kinds of grass that do not accumulate radionuclides. Contaminated cattle dung is a fertilizer, which contributes to the contamination of the private plots. Many local residents use forest wood in their furnaces, which turns them into 'private reactors'. The ashes are then used as soil fertilizer, continuing the cycle. (Ibid.: 43)

The radiation itself and the resources to cope with it are unequally distributed and structure-given. At the same time the doses of radiation are being individualized. Let's assume what is sold in grocery stores has to pass radiation control but what is individually produced does not.

Consumption of free goods from the forest might be less of a want and more of a need. Greater economic vulnerability translates into greater exposure to radiation risks, and this relationship is mediated, paradoxically, by the natural resources of the forest. The relationship among socioeconomic privilege, the use of forests, and risk distribution can be observed on the levels of both communities and individuals. (Ibid.)

This metamorphosis of nature into a civilizational threat produces something new which could be called 'environmental or risk appropriation'. It represents a historically novel devaluation of nature, capital and achievement, while relations of production (ownership) and sometimes even the characteristics of the commodities remain constant.

Picking up once more the question Where does the power go?, the first part of my answer is: the power structure of global risk is centred not only or mainly in the state (as the national perspective suggests) but in the epistemic cultures of experts. As long as we study risk within the institutionalized paradigms of social change, the power and the politics of invisibility remain invisible. Only when viewed through the lens of the theory of metamorphosis do the definitional relations and their historical problematization come into view. New issues and domains of the reality and politics of risk are opened up to analysis. The law, which is nationally conceived, institutionalized and restricted, fails to take account of the susceptibility and vulnerability to risk of other populations and of other nationalities in other parts of the world. Under conditions of cosmopolitization, those practices produce all kinds of contradictions. After the Fukushima accident, for example, we observe the common tactics of this time: the Japanese officials downplay the scope of the catastrophe by withholding data and raising the limits for radiation exposure at least twenty times, including for children.

We need to distinguish between two different power models of risk definition by global epistemic communities of experts: the *nuclear* model, in which the experts are both creator and assessor of the risk they create, and the *climate change* model, in which the climate scientists are experts on side effects.

3   Politics of Invisibility: Nuclear Science

In the nuclear model the experts who define the risks are both creator and assessor of the nuclear risk they create. Their power structure is determined by the power of the nuclear industry and their dense interconnectedness and cooperation with the state bureaucracy. The main consequence is that, whenever they acknowledge the uncontrollable nature of nuclear risks for the affected population or, in the worst case, for all of humanity, they violate *their own* vital interests as well as the interests of the industry and the state.

In the case of political domination, the invention and implementation of democracy introduced the norms of the separation of power. One of the main characteristics of the nuclear power complex is that there is *no* separation of definitional power. Translated into different terms, nuclear expert power is constituted in the unity of the 'executive' and the 'judiciary' of nuclear risks. There is no institutional separation between those who produce and those who diagnose the risks; indeed, questions about this are rejected with reference to the 'scientific rationality' that is the distinguishing feature of expert judgement. In this way, issues are stipulated in a preventive institutional manner: Who has access to the (research) resources for diagnosing risks? To the research questions that are raised or not raised? To the funding and publication of research findings? Who can participate in the discussion? Who is in charge? Who must keep quiet? In the case of nuclear risks, these are ultimately not questions at all, because the experts who produce and diagnose the risk have a global monopoly on definition – both towards states and towards national legal systems.

The definitional power of the Western nuclear industry is globally organized, as summarized by Kuchinskaya and many other studies, 'in the lack of adequate international assistance on the state level, in the international studies that worked hard to find nothing, and in the reports that ignored the research of local scientists and blamed the affected populations' lifestyles or their fear of radiation' (2014: 160).

Taking this into account, surprising facts come into view. The 'risk-nation' that is mainly threatened by the consequences of the nuclear catastrophe is organized as a place and an actor whose objective is to break the power monopoly of international nuclear experts and their organizations.

In committees and conferences in which international and national experts meet, the politics of invisibility is being questioned for the first time by highlighting suppressed facts. Among them is the fact that many doctors have left the allegedly non-hazardous zones and the question both of why the threshold values that regulate which claims to compensation are justified are constantly being changed and of which parts of the population must be evacuated – and so forth. Interestingly, these contradictions are precisely what ignites the opposition of the nations at risk to the appeasing diagnoses of non-hazardousness of the international nuclear experts. 'The Belarusian scientists proposed their own, alternative concept. From their perspective, people could not live where it was not possible to obtain uncontaminated food and where normal life activities had to be limited' (Kuchinskaya 2014: 71–2).

In the transition process from the disintegrating Soviet Union to the independent post-Soviet states a national resistance formed comprising local experts and national politicians. In contrast to the international nuclear expert organizations, they accorded the consequences of Chernobyl the status of a 'national disaster', with all the social and political consequences that follow from this. The resulting conflicts revolved in particular around the claims to compensation for those who lived and worked in the areas thus considered to be contaminated. However, this recognition policy was thwarted by the fact that the ensuing, unmanageable, indeed unimaginable, costs became the central focus. At the same time it was emphasized that the integration of the new, independent state into the Western market economy presupposed that the consequences of Chernobyl had to be eliminated or (rendered) invisible.

The antagonism and disagreement between the powerful international nuclear risk experts and local experts who experience and analyse the complexity of the risks on the ground can be shown by how the still largely unresearched long-term effects of small doses of radioactivity are dealt with. No one really knows the long-term effects of doses of radiation that are classified as tolerable in the short term. The more or less radioactively contaminated areas provided an ideal opportunity for such a study. But attempts to finance corresponding research projects foundered on the resistance of the IAEA and the Western nuclear experts, who argued, among other things, that they had the situation under control with the aid of their satellite-based monitoring systems. It should be noted here that this resistance cannot necessarily be interpreted simply as a defence of power monopolies. Instead, the belief in the rationality of one's own position is precisely what excludes the question of power and thereby supports and protects the position of power. All forms of opposition are denied the attributes of rationality and are dismissed as amateurish, dilettantish and hysterical. 'There is a [globally organized] nuclear lobby there that preaches that nuclear power stations are harmless' (Kuchinskaya 2014: 124). They defend the future of the nuclear industry with all the means of power they have.

4 Politics of Visibility: Climate Science

How does the creating and maintaining of *public visibility* become possible? As we have seen, in the case of nuclear risk, public visibility depends on a (local) expert opposition backed locally and nationally by the most afflicted nation against the global power of the international epistemic community and its organization. It depends on the counter-power of independent experts because of the 'natural' invisibility and because of the need therefore constantly and critically to examine the empirical conditions and the standards of protection. In other words, even those who are most afflicted are dependent on the scientific and administrative means of visibility. Without them they live like fellahs in ancient Egypt.

Most importantly, this implies a 'risk democratization' that democratizes access to and possession of the power relation of definition. Focusing on the power relations of production, we realize that a whole set of modern institutions have been created and implemented in democratic societies to limit the power of capital and to empower the workers – institutions such as the unions, the welfare state, labour laws, etc. Nothing like this has happened on behalf of the relations of definition in world risk society. The norms of accountability are, to say the least, insufficient (especially on the international level). The opposite is true: the nuclear industries and their experts were able to build a global monopoly of definitional power practising most sophisticatedly the politics of invisibility. In the case of nuclear risks, there is no doubt: in order to be effective, advanced risk societies must be democratized. This requires a reform of the relations of definition. Because we either fail to recognize or deny the global risks of radicalized modernity (a persistence of unawareness), the world is a far more dangerous, perhaps fragile, place.

Thus, as we have seen, the socio-economically most vulnerable are those who suffer most from the social construction of ignorance. In more theoretical terms, the relations of definition are subordinated to the relations of production. This signifies the status quo in most risk societies ('risk-CLASS'). But there is no necessity for this subordination. Indeed, as some societies demonstrate, reforms of the relation of definition can be pushed forward while holding still the relations of production. For example, the phasing out of nuclear energy and targeted political development of alternative energy sources in Germany shows that relations of definition can be reformed and democratized while property relations remain constant.

Many of the characteristics of the nuclear epistemic community do not characterize the epistemic community of climate scientists. They are not Janus-like experts who profit from both the creation and the assessment of risk. This fundamental structural dislocation, which seems to be the normal case, has been overthrown. Their diagnosis of global warming arises from the fact that they are 'specialists on side effects'. This is a completely different role and model of expert power.

The power structure of risk in the case of climate scientists is (1) organized in a way that they do not violate their own interests by monitoring and articulating the public risk of global warming. (2) The opposite is true: the politics of public visibility of the also imperceptible climate risk to humanity *increases* their definitional power and social status. (3) Creating and maintaining public visibility by opening and defending spaces specially for those who are most affected by global warming is an important part of their profession-alization. (4) There is a structural independence from those industries which produce the climate risk. Therefore (5) there is a division of power between those who produce the risks and those who assess the risks. (6) Last but not least, their assessment of the 'side effects' that threaten the survival of humanity comes from the natural science, its scientific means, definitional power and authority.

Climate scientists point to or produce a global concern. The global distinction and inequality between those who produce the risks and those who are afflicted and threatened by them becomes visible – against the institutionalized poli-tics and law, which are nationally organized. Thus their politics of visibility is twofold: its goal is to make the invis-ible threat to humanity visible. With that they create a cos-mopolitan outlook which makes the social structure of power and inequality visible.

In contrast to the case of nuclear risk, in the case of climate science the link between the state and the defini-tional power of experts does not exist. There is no corre-sponding political actor for the assessment of global climate risk. And the translation and implementation of effective climate politics in the various national political contexts is met by all kinds of obstacles. This is because the legitimacy of national institutions actually arises from the denial of global warming.

At the same time climate science has redefined the so-called universal as well as the so-called national. Interesting enough, the IPCC is attacked by climate deniers in the name of science. And they are also attacked because they bring in too many 'national interests' or are even criticized as an 'international lobby' – the lobby of model-builders.

The very system of the IPCC stands for a new way of connecting global problems, cosmopolitan science referring to and transcending national boundaries because every nation assesses something about the report. There is a national dimension in it and at the same time there is a new kind of institution – a 'cosmopolitan world parliament of (climate) science', creating a kind of contextual universalism that represents the many different national and local voices and thereby including local and national knowledge.

What can be observed here is that, for all the different topics – forests, sea levels, fisheries, agriculture, transport, cities – totally new institutions have been invented and implemented that have reorganized all the connections between nature and societies.

Climate scientists are caught in the paradox: creating visibility of global warming on the global level implies creating invisibility on the local, the national and the regional level.

5 Outlook

There is a remarkable consensus among the nuclear experts of the Soviet Union and the Western experts of developed democracies. 'A catastrophe whose scale was unimaginable... became *manageable* through a particular dynamic: nonknowledge became crucial to the deployment of authoritative knowledge' (Petryna 2003: 39).

The consensus on maintaining, indeed re-establishing, the relations of definitional power overcomes (or overcame) the ideological and historical boundary separating the communist system of the Soviet Union from the capitalist-democratic system of the United States and Western Europe. However, with a time lag – namely, three years after the Chernobyl nuclear explosion in May 1989 – a political explosion occurred. 'Public awareness of the contamination and the extent of the Soviet cover-up exploded' (Kuchinskaya 2014: 119).

Local scientists criticized the 1986 Soviet report as misinformation. Not only the catastrophe and the aftermath of the catastrophe but also the anticipation of the future catastrophe in the present form a new kind of revolutionary force. In other words: world risk society metamorphoses the notion of the revolution (see later).

Conventionally we understand revolutions as poverty revolutions, taking place in the centre of the political system, often under the ideological and political leadership of middle-class intellectuals promising to realize the basic values of equality and justice (or fighting those values and re-establishing authoritarian structures). To some extent 'the reality in itself' is a natural force of resistance against the politics of invisibility. In world risk society, producers and carriers of risk conflict with each other (and debate among themselves) over relations of definitions that are largely designed for and unchanged from modern nation-state society, which are historically inadequate for global risk society.

Relations of definition become exposed and politicized with each catastrophe that reminds us of the globality of risk society and as the logic of global risks permeates everyday experience. The combination of antiquated national relations of definition and the global politicization of science makes visible the underlying structure of 'organized irresponsibility', as situations in which individuals, organizations and institutions escape accountability for exactly those risks and potential disasters which escape the preponderance of laws and regulations.

In the current era of manufactured uncertainty, the constant threat of a growing array of local risks and mega-hazards opens up sub-political, sub-revolutionary spaces of actions and reinvents scientific and political infrastructure institutions. This is not only about new cosmopolitized spaces of action but about new fields of political action and reforms.

7

Emancipatory Catastrophism: Common Goods as Side Effects of Bads

When one day the history of emancipatory catastrophism is written, it will start not with the issue of global climate risk but with the experiences and horrors of the Second World War, as a significant historical shift in which the emancipatory potential of the risk of global war has led to a series of cosmopolitan institutions, such as the UN, the IMF, the World Bank and, most significantly, the European Union. It was a moment of cosmopolitical metamorphosis. Of course, this is a post-argument. It is *not* to suggest that we need a catastrophe like the Second World War to achieve emancipatory politics. It is the experience of the catastrophe that violates the 'sacred' norms of civilization and humanity and, with that, creates an anthropological shock from which institutional answers become possible and can be institutionalized on the global level, not automatically but through significant cultural and political efforts.

The institutionalized realization of the emancipatory potential of the 'world' catastrophe is subject to enormous resistance. At the same time, it is open for potentially indefinite revision. It is not ahistorical and fixed.

The question today is: Does the unfolding catastrophe of climate change constitute, in the same way as the Second

World War, a potential for emancipatory catastrophism and the implied realization of cosmopolitan institutions?

This chapter elaborates on the *positive* side effects of global risks, such as global climate risk. For those whose ideas are rooted in the paradigm of social change, this question simply does not arise, because this paradigm rules out the idea of global risks and locks the observer into the modern frame of normal risk. This makes risks fit into existing institutional frameworks, which, however, are not only unable to deal with them but actually reproduce and foster them.

Metamorphosis is about the hidden emancipatory side effects of global risk. As chapter 5 demonstrated, the notion of 'risk-class' is about the co-production and co-distribution of goods and bads. This chapter goes a step further and demonstrates how the theory of metamorphosis goes beyond world risk society theory: it is not about the negative side effects of goods but about the positive side effects of bads, such as the moment of cosmopolitical metamorphosis triggered by the Second World War. They produce normative horizons of common goods and replace the national with a cosmopolitan outlook. This is what I call 'emancipatory catastrophism' – but, again, it is a post-argument and not a plea for global catastrophes.

1   How Metamorphosis of the World Works Can be Seen and Analysed through Three Conceptual Lenses

The expectation of global climate risk, despite all the pessimism about the failure of adequate political answers and action, has already invested the postmodern 'everything goes' attitude with a new – if not utopian, then dystopian – meaning. Global risks – such as climate change or the financial crisis – have given us new orientations, new compasses for the twenty-first-century world. We recognize that we have to attach central importance to the dangers that, until now, we have repressed as side effects. Climate change is not climate change; it is at once much more and something different. It is a reformation of modes of thought, of lifestyles and consumer habits, of law, economy, science and politics. Whether presenting climate change as a transformation of human authority over nation; as an issue of climate (in) justice; as concerning the rights of future generations or the relationship between moral rights and climate issues; as a matter of EU politics or international trade; or even as an indication of suicidal capitalism (chapter 8) – all this is about the dramatic power of the unintended, unseen emancipatory side effects of global risk, which have already altered our being in the world, our seeing the world and our doing politics.

Global climate risk could usher in a rebirth of modernity. Haven't climate scientists set a transformation of capitalism in train that is self-destructive and destructive of nature, a transformation that was long overdue, but seemed impossible before? Isn't the agility with which the Chinese today are promoting the boom in the trade in renewable energy sources an example of the co-evolution of the opponent? Thus Western climate sceptics are violating their own economic interest. Perhaps it is reasonable to take all nuclear power plants offline regardless of whether or not they are safer than the Japanese models – this, in any event, solves the problem of the final disposal of spent fuel rods. And, either way, the renewal of solar and wind energy is a meaningful renewal of modernity.

Perhaps the topos of climate change is even a form of mobilization thus far unknown in human history that breaks open a sanctimonious national autistic world with the vision of the impending apocalypse? Could it be, then, that the global climate risk, far from being an apocalyptic catastrophe, can be changed by active (cultural) work and cooperative politics of many actors into a kind of 'emancipatory catastrophe'?

Three conceptual lenses are useful to grasp how the metamorphosis of the world works. First, violation creates the norm (and not the other way around). The anticipation of global catastrophe violates *sacred* (unwritten) norms of human existence and civilization. A violation of sacred values causes, second, an *anthropological shock* and, third, a *social catharsis*. This is how new normative horizons as a frame of social and political action and a cosmopolitized field of activities emerge.

As argued above, this emergence of a compass for the twenty-first century world must not be misunderstood as something that automatically happens or is naturally caused by the event as such. Rather, it is the product of *cultural work*. The confrontation between existing institutions of law and politics and these new normative horizons leads to a permanent process of reformation and counter-reformation – non-linear and open-ended. Metamorphosis is a process in process. We do not know the ending – maybe there is none. And there are lots of complications. First of all, there is powerful resistance. Also there is the paradox of empty promises. Think about human rights agreements: they were ratified by dictators before 1949. Now, the empty promise catches up with them. How, then, does the anthropological shock become law? That is a long, sometimes endless way and story.

2   Hurricane Katrina: How the Normative Horizons of Climate Justice are Globalized

The case of Hurricane Katrina, the anthropological shock it stood for, is a prime example to shed light on the pressing and, so far, neglected question of how the normative horizons of climate justice are actually globalized.

The hidden emancipatory side effects of Hurricane Katrina unfolded when it hit the coast of Louisiana on 29 August 2005. This is manifest in how the literature reflected on the event. Analysing the discourses around Katrina shows a paradigm shift – in fact, a social catharsis – in that two formerly distinct discourses came together: ecological challenges and the history of racism in the US. An excellent illustration is offered by Quincy Thomas Stewart and Rashawn Ray, who use the metaphor of 'race flood', referring to the fact that most people hit by the flooding were black and poor. They argue

> that this natural disaster mirrors a social catastrophe that has affected the lives of Americans since the colonial era – *the race flood*. Just as the hurricane and ensuing flood penetrated the lives of New Orleans residents, the concept of race has penetrated American social institutions such that racial classification shapes the breadth of an individual's social interactions, perspectives and life chances. Race, in many ways, is one of the primary lenses through which Americans view, experience and appraise their social world. (Stewart and Ray 2007: 39)

Until Hurricane Katrina, flooding had not been positioned as an issue of environmental justice – despite the existence of a substantial body of research documenting inequalities and vulnerability to flooding. It took the reflection both in publics and in academia on the devastating but highly uneven 'racial floods' of Hurricane Katrina, bringing back the strong 'Anthropocene' of slavery, institutionalized racism, vulnerability and floods for the substantial community of environmental justice academics and activists in the US, to turn their attention to risk that was seemingly 'natural' but has to be uncovered as essentially social and political. In this way the social birth and career of the cosmopolitan perspective and horizons of justice can be empirically located and studied: 'A small but growing body of literature is now framing flood risks in the US and elsewhere as a question of environmental inequality and injustice (e.g. Bullard and Wright, 2009; Dixon and Ramutsindela, 2006; Ueland and Warf, 2006)' (Walker and Burningham 2011: 217).

It was this social catharsis that led to the emergence of a new normative horizon, namely the global justice frame – i.e., which produced a common good as a side effect of bads. Katrina made clear that climate catastrophe and racial inequality are closely interlinked. This made obvious the inseparable connection between climate change and global social justice. The traumatic experience produces a process of reflection in which things which had not been thought of as being connected are now connected – flooding of cities with racial inequality with questions of global justice.

The social catharsis, however, must not be misunderstood as something that automatically happens and is inherently caused by the event as such. It is the product of carrier groups engaging successfully in 'cultural work', in transformative 'meaning work' of activists in witnessing the distant suffering of others (Kurasawa 2007, 2014). This meaning work is to provide answers to the following questions: What is the nature of the threat? Who are the victims and how do they relate to the publics involved? One of the specifics of global climate risk is that there is no difference between victims and publics (risk to humanity). Who is responsible? And, last but not least, what should the global community and individuals, communities and organizations, wherever they are now, be doing in response?

The cultural work is not simply about the representation of the events as such but about the symbolic environment within which and against which the event is perceived, the imaginations of catastrophe – e.g., as presented in climate publics (chapter 4) or in practices of climate aesthetics (intertwined with scientific and mass media events) and in popular culture (comics, blockbuster movies, sci-fi novels, etc). 'Art practices are actively attending to this "risky" cosmopolitization, giving aesthetics voice and "visuality" to unfolding climate issues and concerns and hereby practicing...an aesthetics of cosmopolitization!' (Thorsen 2014).

In order to produce civic power, these cultural carriers and meaning workers have to construct extra-national events locally in a manner that, despite different languages and histories, often reveals a high level of intertextuality creating the common understandings.

An example of 'transformative work' is provided by Gordon Walker, in his study of how the environmental justice frame has broadened and diversified across topics, contexts and continents.

The spatial-cultural and institutional contexts in which justice claims are being made and justice discourses are being articulated are globalising far beyond the USA to include, for example, South Africa (London 2003), Taiwan (Fan 2006), Australia (Hillman 2006), the UK (Agyeman and Evans 2004), New Zealand (Pearce et al. 2006), Sweden (Chaix et al. 2006), Israel (Omer and Or 2005), and global contexts (Adeola 2000; Newell 2005). (Walker 2009a: 614)

As Gordon Walker argues, globalizing the normative horizons of climate justice can be observed and studied in two ways: horizontally and vertically. Horizontally is, of course, an issue of international networking and *globalization from below.*

The Coalition for Environmental Justice, [for example,] a civic action network of activists, lawyers, and researchers from environmental and human rights organizations in Bulgaria, Czech Republic, Hungary, Macedonia, Romania and Slovakia, was established in 2003 to actively promote an environmental justice frame across Central and Eastern Europe. Network activities included linking up with environmental justice activists in the USA to form a 'Transatlantic Initiative on Environmental Justice' in 2005 (Pellow et al., 2005), and the laying out of an agenda of key issues for Central and Eastern Europe. (Walker 2009b: 361–2)

Looking at the diffusion of the expectations on justice, Walker also distinguishes processes of globalizing *vertically,* which are not disconnected from the horizontal travelling of ideas and meanings across borders. The 'logic' of global risk thereby becomes real, including the positioning of transnational responsibilities for harm in distant locations, connecting global economic and political relations with their local or national environmental consequences. For example, the agenda of the Coalition for Environmental Justice's transnational network in Central and Eastern Europe includes the exporting of risks from richer to poorer countries along a range of country-specific concerns (Steger 2007).

The 'becoming real' of the cosmopolitan perspective can, of course, also be studied through the analysis of the ways in which climate catastrophes are (re)presented in mass-media and digital communication (chapter 8). It is therefore necessary to distinguish between flooding or other catastrophes in specific space and time, on the one hand, and the global risk of climate change as an existential risk to human-kind, on the other. Global risks (like global climate risks) are not the result of any specific catastrophe to others in any specific space and time. Rather, they need to be staged ('socially constructed') as anticipated catastrophes to human-kind *for-us*. The question then becomes, How does the cosmopolitan perspective become 'real' – that is, an across-border reality – 'for-us'?

This, one might speculate, presupposes, for example, the representation of a sum of interconnected *national* perceived tragedies. The 'link' between the interconnected national tragedies could be, for example, mass tourism. Mass tourism being involved in and threatened by climate catastrophes is a way in which the geographical and social distance – the catastrophe 'for-others' – is being metamorphosed into a catastrophe 'for-us' by the social nearness of a catastrophe 'distant from us'. Television, email and satellite telephone enable people to remain in contact with their loved ones and make horrifying images and videos available at the click of a mouse.

3　Outlook: Compass for the Twenty-First Century

Anthropological shocks occur when many populations feel they have been subjected to horrendous events that leave indelible marks on their consciousness, will mark their memories forever, and will change their future in fundamental and irrevocable ways. Anthropological shocks provide a new way of being in the world, seeing the world and doing politics.

From this a social catharsis may emerge, including reflex, reflexivity and reflection. The anthropological shock induces a kind of compulsive collective memory of the fact that past decisions and mistakes are contained in what we find ourselves exposed to; that even the highest degree of institutional reification is nothing but a reification that can be revoked, a borrowed mode of action, which can and must be changed if it leads to self-jeopardization. Global climate risk, but also global financial risk, etc., is discovered in public discourse and reflection as the embodiment of the mistakes of ongoing industrialization and financialization.

Metamorphosis is not a revolution, not a reformation, not something which is intentional, goal-orientated, part of or a result of an ideological struggle (between parties or nations). It is – as I try to show with the case study on climate change – proceeding latently, behind the mental walls of unintended side effects, which are being constructed as 'natural' and 'self-evident' by law (national and international) and scientific knowledge production.

But this is only part of the story; the other part is that the anthropological shock of catastrophe creates a 'cosmopolitan moment'. In this moment of catharsis the mental walls of institutionally constructed side effects are breaking down, and we can empirically study the social fact of how cosmopolitan horizons are emerging and being globalized.

I did not argue in terms of a philosophical-normative cosmopolitanism. I argued that climate change *empirically* produces a basic sense of existential and ethical violation of the sacred, which creates the potential for all sorts of normative expectations and developments – norms, laws, technologies, urban changes, international negotiations, and so on. This is the power of *metamorphosis* towards a cosmopolitan horizon of normative expectations. This is the critical standpoint.

This critical standpoint has to be clarified. It is empirical and normative at the same time. But the normativity of this critical standpoint is very specific. It is about the empirical acceptance (*Geltung*) of 'value relations' (*Wertbeziehungen*, as Max Weber called them). They must not be confused with value judgements using value-loaded terms, sentences and explicit moral languages. They are empirical in the sense that they can be studied from an observer perspective.

The discourse on climate justice has uncovered a number of hurdles – sometimes hurdles of a theoretically troubling kind. One example is that issues of climate justice include future generations who are going to suffer most. Therefore the problem arises as to how to address and apply norms of justice to subjects who do not yet exist and therefore have no voice of their own in making decisions that will dramatically affect their conditions of life? Often those unjustly injured by climate change risk cannot complain to anyone in particular. This, in fact, makes it easy to apply the existing national law system, which excludes the excluded.

At the same time a vision for climate justice has to recognize sooner or later that persistence of colonial historical patterns and the dense intimacy of their linkages and dynamics with law's constitution of both its 'subject' (the legal actor) and 'the environment'. The problem of climate justice discovers links between the colonial foundations of international law and the philosophical foundations of the Western juridical imaginary. What is at stake here empirically, and therefore normatively, is a mode of violation aimed at the living order itself. But at the same time we have to be careful not to confuse the difference between 'dependency' and 'cosmopolitization': problematizing climate injustice by pointing at those individuals, communities and nations who have been on the wrong side of colonial history, who have suffered and continue to suffer, is in itself an indication that cosmopolitization enforced by global climate risk creates a normative horizon and reflexivity about exactly that fact. More than that, it creates again (as a fact) the expectation (sometimes even the conviction) that a reformation of institutions (law, politics, economy, technological practices, consumption and lifestyles) is now urgent, morally imperative and politically possible – even if it fails at conferences and in politics.

I have tried to show that, on the basis of the empirical globalization of this critical standpoint, we are able to criticize what one might call the national (and transnational) domestication of climate change, the post-political consensus around 'green economy', technological innovations, etc. This is where things become a matter of political economy, and, from a cosmopolitan perspective intrinsically connected to climate change, we can include and mobilize the new global geographies which do not respect the post-political European 'consensus' in any way. This is also a key point in terms of the *metamorphosis* of international power relations (chapter 10).

Seen in this way, climate change risk is far more than a problem of measures of carbon dioxide and the production of pollution. Nor does it signal only a crisis of human self-understanding. More than that, global climate risk signals new ways of being, looking, hearing and acting in the world – highly ambivalent, open-ended, without any foreseeable outcome.

Metamorphosis then also means that the past is reproblematized through the imagination of a threatening future. Norms and imperatives that guided decisions in the past are re-evaluated through the imagination of a threatening future. From that follow alternative ideas for capitalism, law, consumerism, science (e.g., the IPCC), etc.

It even includes a self-critical approach to everyday norm-creation in the mode of dogmatism. In the technocratic version of environmental politics, carbon emissions become the measure of all things. How much carbon does an electric as opposed to a manual toothbrush produce? From now on, even divorce is answerable not only before God but also before the environment. Why? Singles consume much more energy and other natural resources than couples living together in a household.

As a result, a compass for the twenty-first century arises. Yet, in contrast to the Second World War, in the case of the global risk of climate change, it is an open question where this compass leads us. There is an enormous discrepancy between normative expectations and political action.

8

Public Bads: Politics of Visibility

The relationship between 'communication' and 'world' is central to the social theory of modernity. Although it often lacks recognition, Karl Jaspers's key contribution to our understanding of modernity was the invention of the concept of *Weltkommunikation*. It was then Luhmann (1995) and Habermas (1987) who took the notions of communication and communicative action, respectively, as the key notions in their theories of modern society. In my theory of metamorphosis, communication plays a key role too but in a fundamentally different way – in fact, it is thought of and conceptualized in a fundamentally different way, namely, through the perspective of metamorphosis. Applied to the theory of modern society, it means that this is about the metamorphosis of modern society and politics. There is no metamorphosis without communication: communication about metamorphosis is constitutive of metamorphosis.

So far I have explored this epochal change of horizons through the metamorphosis of social inequalities: from risk to risk-class, risk-nation, risk-region. I have further analysed the metamorphosis of power: power relations of definition as opposed to power relations of production. Finally, I have discussed the sociology of metamorphosis through the example of the relationship between catastrophe and emancipatory catastrophe. In order to explore the significance of communication for the metamorphosis of the world, I will not develop a universalistic theory of the communicative constitution of the world. Rather, I will introduce the middle-range concept 'public bads' as a means of cosmopolitan theorizing. I will do this in two steps: first, I will introduce the concept of 'landscapes of communication' and explore their metamorphosis; second, I will explore the concept of 'public bads'.

1 New Landscapes of Communication

In times of digital communication, world risk society accounts for an important structural dynamic through which global risks create new forms of 'communities'. To understand this structural dynamic means to understand the metamorphosis of modern society in the digital age.

Global risks (climate change and the financial crisis) have the power to change society and politics, but only in the medium of public communication. Global risks per se are invisible. It is only through mediated images that they acquire the power to break through this invisibility. Large-scale disasters are occurring everywhere, but they unfold their emancipatory potential only with the power of the public images that create a global public sphere, a categorically different kind of public than the one trapped in the national view. What we can observe is an interaction: global risks create globalized publics, and globalized publics make global risks visible and political.

The French media theorist Paul Virilio encapsulated this power of images in the phrase 'Images are ammunition, cameras are weapons'. Global risks are turning into battlegrounds of visual globalization. It is not the catastrophic events but the globalized images of these events that trigger the anthropological shock, which, filtered, channelled, dramatized or trivialized in the diversity of the old and new media, can create a social catharsis and provide the normative framework for an ethics of 'never again'.

Once more, it is not the images that accomplish this, but the globally mediatized and commented-upon images that make the millionfold visualization possible. Whether it is the despair of a Palestinian father holding his dying son in his arms in the Israeli–Palestinian conflict or the brutality with which the IS performs and celebrates the beheading of Western hostages before the eyes of the world, pictures make their way around the globe and are powerful political instruments. In such symbolically condensed images, historical conflicts and political struggles are intensified, translated, personalized, but also instrumentalized, truncated, simplified and falsified. It is not the catastrophe as such but the globalized and figurative communication about the catastrophe that first releases the emotion, and perhaps also the identification with the suffering of others, that triggers an anthropological shock capable of abruptly changing the political landscape.

The world of the mass media has long been, and today is still largely, a world of nations. Indeed, as first the German philosopher Hegel and later the historical sociologist Benedict Anderson (2006) demonstrated in pioneering studies, the invention of printing contributed in essential ways to the production and reproduction of national consciousness, and hence of the nation as an 'imagined community'. In the meantime, the old mass media (newspapers, radio and television) have become increasingly open to global events. To this are added the diverse, rapidly evolving new communication media (the internet, Facebook, social media, smartphones, Skype, etc.). This development has also given rise to communication networks and flows that extend across borders and mark the end of the national communications systems.

Global communication (and hence also in a different sense world history) is only just beginning. Until now, global communication did not exist; there was only an aggregate of national forms of communication. Even where these were interlinked, the selection of events and the way they were covered reinforced the underlying national or local horizon.

Today there is no longer any outside or inside. The reference frame of communication is no longer this or that nation – rather, the communication situation is one of humanity as a whole (which should not be confused with the normative horizon of 'global public opinion' – more on that later). New landscapes of global communication are taking shape in which the particular, fragmented and globalized horizons of Facebook communication overlap, interpenetrate and mingle with national public arenas.

2 Public Bads

It is against the background of these new communication landscapes that I am introducing the time-diagnostic notion 'public bads' as a focus for cosmopolitan theorizing and research. By 'public bads' I refer to the constitutive connection between global bads and global publics. There are no bads – global risks – without global publics. Simultaneously, global risks create global publics and thereby reconfigure the national landscape of public communication. The notion of public bads focuses on the conceptual intersection of anthropological shock, side effects and future risk imaginations. Through the lens of global public bads, the complex filigree architecture of the metamorphosis of the world of media, communication and publics can be studied.

Is 'public bads' merely another name for 'global risk'? No, this is not the case. Rather, it encapsulates what the concept of global risk conceals – namely, that global communication and the global public arena are constitutive of global risks.

The notion of public bads combines different social theory trajectories. In part it refers to the notion of side effects and global risk. For side effects, there is John Dewey's work on *The Public and its Problems* (1954). According to Dewey, the public arises not from political decision-making but from the experience of negative side effects of the actions of others. He argues – to use my terms – that bads produce publics and thereby enforce the search for a new institutional order.

The notion of public bads is further characterized by three elements: interconnectedness; public – that is, reflexive – interconnectedness; and the ambivalence of reflexive inter-connectedness created by bads. In this respect, it is related to Manuel Castells's *Network Society* (1996). Yet a major difference is that you can log out (at least in principle) of the connections of network society, whereas you cannot log out of public bads. There is no escape. The future is a curse and a blessing of communicative coexistence of everyone with everyone.

Progress publicness and risk publicness

In this context I suggest that we distinguish between two forms of communication and their respective public dimensions: on the one hand, 'progress publicness' and, on the other, 'side-effects publicness' or 'risk publicness'. Progress publicness is connected to the fact that in all democratic societies there is a public discussion about the future of modernity. This focuses on the production and distribution of goods in the national or the international context and on the resulting social and political dynamics. The questions and conflicts surrounding the production and distribution of goods, and the attendant social and political dynamics of power and class and democratic forms of government, are essentially geared towards promoting 'progress' while down-playing the associated side effects (bads). Accordingly, the discussion turns on goals, decisions, political ideologies, etc., and conflicts over different conceptions of the future are conducted in these nationally organized public spheres strug-gling over progress. The mode of this kind of nationally organized public dimension of media power is exclusive. It is deliberately produced; it can be permitted, suppressed, and so forth.

Side-effects or risk publicness, which centres on the pro-duction and distribution of bads (risks), develops in competi-tion and conflict with this. Here the metamorphosis of communication and the public dimension begins to unfold.

Side-effects publicness focuses on the culturally perceived violations of nationally organized progress that are widely ignored in the mainstream public. This is not just a change of subject but also a change in the form of publicness. Side-effects publicness cannot be easily controlled by the powerful. It takes a stand against the risk-oblivious coalition of advocates of progress composed of experts, industry, government, political parties and the mass-media establishment. Side-effects publics arise without being planned, in opposition to the hegemonic discourse on progress, and they are difficult to control. The thematization of side effects marks a second stage in the metamorphosis of publicness. What I called 'emancipatory catastrophism' can arise: the normative horizon of a shared destiny takes shape in the existential threat to humanity. What previously counted as 'bads' are now regarded as 'goods'. A spectacular metamorphosis takes place, which can be captured by Friedrich Nietzsche's words 'the revaluation of values'. It is a radical form of metamorphosis which is not only becoming apparent today in the area of global climate risk but also has historical precursors in other subject areas.

When the struggle for women's emancipation began, feminists were decried and ridiculed as 'ugly blue stockings', 'sexless, man-eating Amazons' who violated the decrees of God and nature. Nowadays, by contrast, a reappraisal of values can be observed in the West at least: anyone who publicly opposes gender equality has lost the political game. Indeed, more than that, we now encounter varieties of opportunistic feminism. The call for equal rights serves as an argument and an instrument to erect barriers against immigration – a case of the politics of the lesser evil.

This reassessment of values of 'bads' into 'goods' is not occurring suddenly, overnight, in a linear way and from the top down. It involves protracted conflicts that can go on for many years, decades, even centuries. These processes are marked by phases of stagnation and retrogression and are 'path dependent', which means that they do not unfold uniformly and simultaneously but are tied into different historical and cultural contexts, and changing social and political actors try to influence them at the national and international level.

Another example of metamorphosis, in the sense of the revaluation of 'bads', is shown by German debates over migration. In Germany immigrants were long regarded as a threat to German national identity. Contrary to this, protracted struggles have increasingly led to the view that immigration and immigrants are necessary given Germany's future as an ageing society with a low birth rate. In this case, the normative framework remains unchanged: according to the one argument, Germany's future is threatened by immigration; according to the other, Germany's future is jeopardized if there is no immigration. But what is common is that, in both cases, the concern is to secure the future of Germany. Here revaluation of values means revaluation of means.

In the case of immigration, therefore, metamorphosis is taking place while the normative frame of reference remains constant. Here metamorphosis means that Germans' picture of Germany anchored in the law is also undergoing metamorphosis. This, in turn, is also taking place more or less (non-)simultaneously from the actor and observer perspectives. The revaluation of women's emancipation from a 'bad' to a 'good', however, involves a change of horizon. Women's liberation gave rise to 'bads' because the emancipation of women was seen as contrary to nature and to God. Similarly, in the European context, the dominant role of religion and the assumption of anthropological constants that defined the normative horizon of reference had first to be 'overturned'. This religious and anthropological horizon became disenchanted and was replaced by the normative horizon of universal human rights and the principles of equality and justice. Within this horizon, 'bads' undergo re-evaluation and are transformed into 'goods' which henceforth cannot be questioned with impunity.

These examples show that, in the social and political domain, we are always dealing with forms of incomplete metamorphosis. Thus, it never comes to what is known in biology as a complete metamorphosis – that is, change from one fixed state to another final fixed state.

This incompleteness of metamorphosis can take different forms. We can see this in the metamorphosis associated with the perception or recognition of global risks. Here metamorphosis, as explained above, can be described in three consecutive stages: first, self-contained, nationally organized 'goods' publicness; second, the hegemonic discourse of progress is subverted and called into question by the 'bads' publicness that is difficult to control; then a third type of publicness can develop whose key feature is that environmental 'bads' mutate into economic and political 'goods'. What the slogan 'black is beautiful' encapsulates about metamorphosis through revaluation could be expressed in the context of world risk society as follows: sustainability is beautiful, an ecological style of living is beautiful, critique of growth is beautiful, critique of capitalism is beautiful.

This metamorphosis perspective now includes the revaluation of 'bads' into 'goods', not only in the landscapes of digital communication but also in the established mass media, which are still nationally organized and reflect matters of national relevance and national priorities. On the one hand, the global is refracted in the national horizon of relevance. But the presence of multiple current disasters also gives rise to global publicness within the national media. However, this intrusion of global public opinion – this internal metamorphosis of national public spheres – is in turn produced through 'bads': through the craving of the media for catastrophes (tsunamis, Fukushima, forced marriages) or the controversy over the claim that other religions, in particular Muslims and Jews, mistreat boys when they circumcise them for religious reasons. Another 'bad', the euro crisis, has even turned the euro-critical UK into a European public forum on all television channels and all newspaper sectors.

At the same time, the rapidly evolving, new technological variants of digital communication are transforming the concept of the public. Consumers of news are becoming producers of news. National borders and topics are becoming less important. New communication landscapes are emerging – fragmented, individualized and simultaneously spreading out into networks in which the power of the communication media is broken. In the process, key concepts such as 'participation', 'interest' and 'integration', which were assumed to be invariant within the perspective of social change, are changing.

Again, what does metamorphosis mean in this context? For one thing, *categorical metamorphosis*: here the focus is on the concept of public bads (see above). Second, *institutional metamorphosis*: here the competition or overlap or interpenetration of the 'old' mass media (national, monopolistic) with the new digital means of communication (fragmented, individualized, globalized) can be observed. Finally, *normative metamorphosis*: here it is a matter of how 'goods' are metamorphosed into 'bads' and 'bads' into 'goods' and, hence, in turn, how the reporting of catastrophe in at least incipiently involuntary, and for the most part unreflected (but perhaps also often conscious), ways becomes the medium of emancipatory catastrophism.

3   Digital Construction of the World

The digital construction of the world is about the digital metamorphosis of the world. What this means is that every human action, every machine, produces data. We enter a *terra incognita*. This does not mean that everything is new (there is no new heaven or earth), but it does signify a Copernicianian Turn 2.0. There are seven aspects.

1 Digital communication metamorphoses the classical notion of *Öffentlichkeit* ('publicness'). Things which used to be subject to classical publicness are now negotiated: disputes take place between civil and uncivil movements, between journalists and politicians; floodings or terrorist attacks are globally discussed and judged; parents and police search for missing children. The advertisement for products

reflects public opinion and vice versa. Digital communication has become the historical space for public communication. In the past it used to be particular territorial spaces, such as streets, places, churches. The advantages of the digital space are evident: groups can organize without moving physically, costs are low, the exchange happens in real time, physical violence is ruled out. In this sense protest and participation in the web are possible.

However, these possibilities are significantly different from democratic participation. They are shaped and made by market players. Digital communication is in the hands of big transnational corporations. Hence, sovereignty over public debate is occupied by the power of private corporations. This is also true for the technological infrastructure. How long can democracies survive this privatization of public opinion? Will the market economy turn out to be the better plebiscite?

At the same time we observe a move of public regulation towards sub-political actors, as we see with the attempt and decision of corporations such as Facebook and Twitter to ban the public spread of terrorist videos. There are two opposing movements: on the one side, there is the demand for public, state regulation of the web; on the other, it is the web that holds the potential to escape the constraints of the state.

2 Digital communication does not replace the old models of *Öffentlichkeit* (publicness), but we find a *distinct enmeshment of the old and the new*. The classical model of mass media is the ancient theatre. There is a stage in front of which gathers the audience. This distinguishes between the active role of the speaker and the passive role of the audience. This distinction is no longer valid for digital communication. Everybody is 'speaker' and 'audience' at the same time. Even if the consumption of mass media remains

high or increases, the metamorphosis of the world takes place behind this assumed stability simply because there is no distinction any more between online and offline. Digital media have become part of the everyday (see Moore and Selchow 2012). Take, for instance, the ways of interaction and communication in today's schools.

Today the exchange within a school class involving teachers, homework, peers, etc., takes place to a considerable extent within the digital sphere. Instagram and Snapshot are used to share personal experiences with peers, WhatsApp is the source for information about what kind of homework needs to be handed in and when, and YouTube is used to stage and share personal talents with the world, from playing the guitar to gaming. It is easy to overlook the metamorphosis in this because it appears that this is communication as usual, just through different means. Yet, the foundation of this supposedly ordinary communication is not in Munich or Paris, but in Palo Alto and Los Angeles. The servers which store the children's interactions are based in South California or the Arctic Circle. That there is a metamorphosis here, in that the communication is internal to the class but actually not local, becomes apparent only when problems occur – i.e., when the servers are hacked. Only then do the actors themselves understand that the classroom and the circle of friends is already technologically cosmopolitized. The children demonstrate where this is heading. They intuitively stage themselves, their identities and their ideas in the world. Early in school there arises an 'identity game' – that is, a competition about the recognition in the world, about whose influence is the most far reaching, manifest in clicks, 'likes', 'friends', etc.

3 Related to this is a new *unimaginability* of data and numbers. Digital communication stands for the

systematic production and consumption of data to an extent that is no longer imaginable. The national conception of the world follows the model of Russian dolls. The world is imagined as the biggest and universal unit which can be cut down into smaller aggregates. The political world consists of a set of nation-states, the economic world consists of free trade zones, the economy consists of markets, which are ordered by target groups. This kind of thinking and collecting data in containers misses the point of the digital world. These quantities are not countable. They are estimated. They belong to the world of general statistics, which is about size, not about details. The overwhelming quantity of data is in principle nothing new. The moment of metamorphosis arises in a world shaped by the logic of risk and pre-emption, in which these kinds of 'big data' are used as the basis for action to 'improve', such as through the killing of potential terrorists through so-called signature strikes.

4 Whereas existing societies are national, digital communication produces, it seems, a world society. Yet this is wrong. It produces *indefinite numbers of 'world societies'*. This means it produces a reality of social relations which do not work according to the classical logics of *Öffentlichkeit* and 'society'. The digital metamorphosis disturbs or destroys the existing notions of society and publicness. At the same time it produces new notions of society and publicness: the global others are here in our midst and we are simultaneously elsewhere. The point is that this is not the product of force but the precondition of the digital age.

5 The world becomes *individualized* and *fragmented*. The individual – the 'undividable' – becomes the point of reference and, at the same time, no longer matters. It sinks in the unimaginable amount of data. Individualization is the process by which the primary

unit of social and political action is no longer an aggregate or collective identity but becomes restricted to individual persons – the paradigm shift from 'we' to 'I'. As such, it is not to be misunderstood as the neoliberal ideology of individualism.

At the same time individualization and cosmo-politization constitute opposing moments in digital communication. On the one side, digital communi-cation forces individuals to rely on themselves, because it undermines the matrix of given collective identities. On the other side, it forces them to use the resources that the cosmopolitan spaces of action hold.

6 In order to capture the metamorphosis of digital communication, the concept of *meme* is central. Meme refers to a change of perspective, away from the communicating actors and towards communicative content and messages. This perspective is essentially not national because meme does not follow national borders. Yet, the paths of meme are not accidental; they are shaped by characteristics, such as commu-nities, professional affiliations, languages, cultures, risk perceptions.

7 The data produced in digital communication are not simply data but *reflexive data*. Digital communication constantly produces data and a kind of organized reflexivity. In order to understand what this means, we need to distinguish between the perspective of participants and the perspective of observation. The relationship between the perspective of participants and the perspective of observation is one that is shaped by the fact that the communicative actors do not realize that they are observable and observed. This means that there is a communicative situation which seems to be closed for the actors themselves when looked at from the inside, but it is open for all kinds of observations if looked at from the outside. This leads to a 'filter bubble' situation (Pariser 2011),

in which the individual is caught in a digital world that is tailored according to their own preferences and habits.

4 Outlook: Cosmopolitan Data

The above has consequences for what we mean by data. So far, the social sciences have produced data which follow the principles of representativeness and aggregation as the central basis for social scientific objectivity. The exploration of the digital metamorphosis of the world cannot be the slave of these principles. Digital communication has to be understood as the permanent production of non-representative and non-aggregative data *by the actors themselves* and not by social scientists. This basic fact implies an epistemological shift.

What we have with digital communication are data which *constitute the reality* of cosmopolitization. They *produce* cosmopolitization; they do *not* simply *represent* it. They are socially and politically meaningful. This insight is intriguing because, taking up Moore and Selchow's point, the internet is then not only a space of action or a tool to organize, communicate and exchange but 'a process of becoming' (Moore and Selchow 2012: 36) – it is a process of becoming a cosmopolitized world. Hence, the process of cosmopolitization in its epistemological status cannot only be represented through indices, indicators and operational definitions but it can be observed as a process of reality.

To be clear, from this perspective, the becoming of a cosmopolitized world is not a hidden process for which we need complex ways of making it visible, but it is as a process visible in itself. The process and the observation of the process are inherently connected.

Here we have access to the reality of cosmopolitization around various issues which have been discussed so far through a different perspective – e.g., the question as to how transnational communities (of 'migrants') evolve. The same goes for the rise of 'world families' (Beck and Beck-Gernsheim 2014), as well as the power of global risk producing cosmopolitan communities (of world cities). To sum up, we have to distinguish between the notion 'representative, aggregate data' and the notion 'cosmopolitan data'. The latter means data which produce the cosmopolitization of the world (there may be other meanings of this term as well).

'Cosmopolitan data' are not cosmopolitan per se but become apparent as cosmopolitan from a cosmopolitan perspective. Of course, digital communication can be analysed from a mainstream perspective, but it is only through the cosmopolitan perspective that the digital metamorphosis becomes apparent.

On the one side, the new situation of permanent data production opens new perspectives. On the other side, it brings up the problem that the methodological assessment shifts from how data are produced to how data are used and interpreted.

At the same time this data production opens access to new objects of analysis, such as communicative streams, patterns of interaction and mobility on a world scale. It opens up the possibility to study cosmopolitan patterns of relationships and to observe how 'cosmopolitan solidarity' develops, for example, around locally experienced climate catastrophes and perceived climate risks. It enables and invites us to study not only 'cosmopolitan moments', such as the occupations of squares across Europe and the world, but the potential manifestation and thickening of cosmopolitan social structures.

9

Digital Risk: The Failure of Functioning Institutions

Metamorphosis in the face of global risk produces a gulf between expectations and perceived problems, on the one hand, and existing institutions, on the other. Existing institutions might work perfectly within the old frame of reference. Yet within the new frame of reference they fail. Hence, a key characteristic of metamorphosis is that institutions simultaneously work and fail. For an illustration I suggest two empirical arguments: first, the 'digital freedom risk', which refers to the PRISM surveillance program; and, second, the digital metamorphosis of society, of intersubjectivity and subjectivity.

1 Digital Freedom Risk

The PRISM scandal has opened up a new chapter in world risk society. In the past decades we have encountered a series of global public risks, including the risks posed by climate change, nuclear energy, finances and terrorism – and now we face the global digital freedom risk.

Whereas the accidents in the reactors of Chernobyl and later in Fukushima triggered a public debate about the risk of nuclear power, the discussion about the digital freedom risk was not triggered by a catastrophe in a traditional sense.

Rather, it was triggered by the mismatch between the perceived and the actual reality of freedom and data in contemporary (Western) societies that became apparent through the revelations of Edward Snowden. The real catastrophe would actually be an unseen hegemonic control on a global scale. The more complete and total the global control of information is, the more it disappears from people's awareness and is rendered invisible. Hence the distinct nature of the digital risk and the paradox that is implied in it: the closer we come to the catastrophe – i.e., the global hegemonic control of data – the less visible it is. We have become aware of the potential catastrophe only because a single private contractor working for the CIA applied the means of information control in order to tell the world about the global digital risk. Thus we are faced with a complete inversion of the situation.

In this sense our awareness of the global digital risk is extremely fragile, because, unlike the other global risks, this risk does not focus on, result from or refer to a catastrophe which is physical and real in space and time. Rather – and unexpectedly – it interferes with something we have taken for granted – i.e., our capacity to control personal information. But then the mere visibility of the issue triggers resistance.

Let us try to explain the phenomenon in a different way. First of all there are some features that all global risks have in common. In one way or another they all bring home to us the global interconnectedness in our everyday lives. These risks are all global in a particular sense – i.e., we are not dealing with spatially, temporally or socially restricted accidents, but with spatially, temporally and socially delimited catastrophes. And they are all collateral effects of successful modernization, which questions retrospectively the institutions that have pushed modernization so far. In terms of the digital freedom risk, this includes the failure of the capacity of the nation-state to exercise democratic control or the failure of the calculation of probabilities (for insurance protection, etc.). Furthermore, all these global risks are perceived differently in different parts of the world. We are faced with a 'clash of risk cultures', to use a variation of Huntington's phrase. We are also faced with an inflation of existential catastrophes and with one catastrophe threatening to outdo the other: the financial risk 'trumps' the climate risk; and terrorism 'trumps' the violation of digital freedom. This, by the way, is one of the main barriers to global public recognition of the global risk of freedom, which, therefore, has only partially become a subject of public intervention.

How we assess the risk posed by the violation of freedom rights differs from our assessment of all other global risks. The freedom risk poses an immaterial threat. It is not a threat to life (terrorism), to the survival of mankind (like climate change risk or nuclear risk) or to property (financial risk). The violation of our freedom does not hurt. We neither feel it, nor do we suffer a disease, a flood, a lack of opportunities to find a job or a loss of money. Freedom dies without human beings being physically hurt. The digital freedom risk threatens 'only' some of the major achievements of modern civilization: personal freedom and autonomy, privacy, and the basic institutions of democracy and law, which are all based on the nation-state.

Seen like this, the real catastrophe is when the catastrophe disappears and becomes invisible, because the control exercised becomes an increasingly perfect one. This happens to the extent to which our reaction in view of the imminent death of freedom remains an exclusively technical and individual one. In this sense, the perception of the freedom risk is the most fragile among the global risks we have experienced so far.

The ongoing process of social catharsis and the worldwide reaction raised a normative horizon centred on human rights issues with regard to mass surveillance: on the one hand, the right of every person to protect their private life; on the other, the duty of states to protect personal freedom, including personal data. The right to protect privacy combined with the duty of data protection is the supreme international human right. It is found in the UN Universal Declaration of Human Rights 1948 (Article 12), and its legal form is found in the UN International Covenant on Civil and Political Rights 1966 (Article 17,1). These rights imply that personal data belong to the citizen, not to the state or private businesses.

The latter principle is threatened today. Yet the acknowledgement of this fact is a rather fragile one. After all, which powerful actor is interested in ensuring that people continue to be aware of this risk, thus pushing the public towards political action? The first actor that comes to mind is the democratic state. But, alas, this would be like asking the fox to look after the chickens. Because it is the state itself, in collaboration with the digital entrepreneurs, that has established the hegemony over data in order to optimize its key interest in national and international security. The extensive enmeshment of private and public resources of control in this field means that we are moving in the direction not of a 'world state', as many anticipated, but of an anonymous digital central power that controls the private behind a democratic façade.

We tend to say that a new digital empire is emerging. But none of the historical empires we know – neither the Greek, nor the Persian, nor the Roman Empire – was characterized by the features that mark the digital empire of our times. The digital empire is based on characteristics of modernity which we have not yet truly reflected. It does not rely on military violence, nor does it attempt to integrate distant zones politically and culturally into its own realm. However, it exercises extensive and intensive, profound and far-reaching control that ultimately pushes any individual preference and deficit into the open – we are all becoming transparent. The traditional concept of empire, however, does not cover this type of control. In addition, there is an important ambivalence: we provide major tools of control, but the digital control we exercise is extremely vulnerable. The empire of control has not been threatened by a military power, by a rebellion or revolution, or by war, but by a single and courageous individual. A thirty-year-old secret service expert has threatened to topple it by turning the information system against itself. The fact that this kind of control seems unfeasible and the fact that it is much more vulnerable than we imagine are two sides of the same coin.

Who, then, could counter this move towards an anonymous digital central power? Could it be the constitutional rights that are guaranteed by democratic institutions, such as parliaments and courts? Alas, in Germany, Article 10 stipulates that postal and telecommunication secrecy is sacrosanct. That sounds like a phrase from a world long gone and by no means fits the communication and control options provided by a globalized world. Europe provides excellent supervisory agencies – a whole range of institutions which try to assert fundamental rights against their powerful opponents – e.g., the European Court of Justice, data protection officers, and parliaments.

And this is exactly how working institutions fail. Given that they are designed from within a national logic, they are not equipped for the cosmopolitan reality. This applies, by the way, to all global risks: answers based on the national outlook and the political and legal instruments offered by our institutions can no longer meet the challenges posed by the global risk society today.

The individual can, indeed, resist the seemingly hyperperfect system, which is an opportunity that no empire has ever offered before. If digital freedom is endangered, the brave can resort to counter-power, to non-compliance on the job. One of the key questions, therefore, is whether we shouldn't oblige the major digital companies to implement legally a whistleblower union and, in particular, the duty of resistance in one's profession, maybe first on a national scale and subsequently at the European level, etc.

2   Digital Metamorphosis of Society, Intersubjectivity and Subjectivity

Everybody talks about digital revolution and the potential it holds. Digital metamorphosis is essentially different from digital revolution. Digital revolution describes a mainly technologically determined social change that captures the increasing degree of interconnectedness and global exchange. The notion of revolution suggests that change is intentional, linear and progressive. As such, it comes close to an ideology according to which development means to have an internet connection (e.g., Slater 2013).

Digital metamorphosis, on the contrary, is about the non-intentional, often unseen side effects, which create metamorphosed subjects – i.e., digital humans. While digital revolution still implies the clear distinction between the online and the offline, digital metamorphosis is about the essential enmeshment of the online and the offline (e.g., Moore and Selchow 2012). It is about digital humans, whose metamorphosed existence questions traditional categories, such as status, social identity, collectivity and individualization. The status of a person is no longer defined mainly by their position in the hierarchy of occupation but, for instance, by the number of Facebook 'friends', where the category of 'friend' itself has been metamorphosed into something that is not necessarily about acquaintance. As such, digital metamorphosis takes place not where one would expect it but in unexpected places.

The emancipatory side effect of global risk, which is produced here, is the expectation of digital humanism, at the heart of which is the demand that the right of data protection and digital freedom is a global human right, which must prevail like any other human right.

The Snowden revelations regarding mass surveillance exemplify another 'emancipatory catastrophe'. On the one hand, they are triggering an anthropological shock by revealing that, and how, democracies are becoming metamorphosed insidiously and imperceptibly into totalitarian regimes. This process of the metamorphosis of democracy can produce a new form of totalitarian control behind the façades of functioning democracy and the rule of law. On the other hand, this shock, and the very substantial political repercussions through 2013 and into 2015, gave rise to a social catharsis, raising profound normative and legal questions. More than that, a normative horizon has been created which challenges the existing practices of totalizing surveillance by a powerful coalition between states and businesses. This happens against the background that in liberal and advanced neoliberal societies, such as the US or the UK, much of society's (future) wellbeing and progress is built on the idea that the private sector is a key driver. Given the dogma of good governance and the managerial discourse that has come to shape global politics and the politics of global institutions, public problems are naturally dealt with these days through public–private partnerships and through the devolving of more responsibilities onto the individual. Increasingly and inevitably, these strategies build on digital data. There is much hope that big data analysis could solve health problems; data crowdsourcing is regularly used to deal with crisis situations; the funding of (formerly public) activities, such as arts projects, through crowdfunding has become as common as the 'public' struggle against obesity or smoking through social gaming. But these details imply the metamorphosis of established understandings of the character and legitimacy of a whole set of institutions constituting First Modernity's national-international order: long-term metamorphosis in the politics of states, in international relations, and in the institutions and norms established in relation to democratic procedures, the rule of law, relations between state and civil society, relations between public policy and private economic interests, the acceptability of cultural norms and, last but not least, even the notion of subjectivity.

The practices of large-scale surveillance by the National Security Agency, Google, etc., must thus be understood not as a scandal which will soon pass but as a side effect of the success in creating a digital modernity, which is inevitably a modernity in which the private and the public sector and the individual are strangely enmeshed – hence, metamorphosed.

There is a new *digital intelligentsia*, a new transnational digital class, using digital cosmopolitization as a power resource for reshaping the world. These epistemological communities of experts challenge both the nation-state and the citizen. On the other side, individuals are the constant producers of the data oceans. The production of data happens consciously and willingly, such as via social media sites, but also unconsciously, routinely and implicitly through the everyday use of personal devices, such as mobile phones, and the surveillance systems that are built into contemporary environments, such as swipe cards, electronic bus tickets, etc.

Digital being in the world, the digital seeing of the world and the digital imagining and doing of politics is by no means a fate, a necessity, a new kind of 'law of history', which everybody has to accept. The opposite is the case: it is a form and process of metamorphosis that is in the process of replacing one frame of reference with another frame of reference, which so far is mainly unknown or foggy.

3   Outlook

This chapter has discussed a figure of metamorphosis in which political and social order dissolves and a different one arises. This was illustrated with the help of the PRISM case. Metamorphosis becomes apparent in the crossing of four unconventional 'revolutions'.

First of all, digital metamorphosis, in contrast to *digital revolution*, is about the metamorphosis of modes of existence: social closeness is being decoupled from geographical closeness; the distinction between fiction and reality is becoming blurred; and modes of (un)controllability by the nation-state, together with the contradiction of being uncontrollable and controllable at the same time, are beginning to appear.

Second, 'collect it all' is the defining revolutionary principle practised by the NSA, thereby overthrowing the constitutional principles of liberty. 'Collect it all' – that was the point of no return for the surveillance state. Rather than look for a single needle in the haystack, the approach was 'Let's collect the whole haystack' – in the words of one former senior US intelligence official who tracked the plan's implementation. It led to a surveillance establishment that was globally out of control. 'Collect it all' was one of the institutionalized procedures of totalitarianism from inside the democratic system.

Snowden going public was the third revolutionary act; it was revolutionary because it made the invisible visible.

Finally, the cosmopolitan perspective on the digital risk opens up the horizon for alternative action. Those new options are cosmopolitan because they cross and connect actors beyond national, religious, ethnic and class borders. To give an example: the fight is not against the USA but against the NSA and in favour of the American Constitution – hoping that the constitutional tradition is too great to fail in this situation. Therefore one of the options Snowden offers is the trust in the American constitutional system and its judges to make the historical decision against the digital threat to liberty. Remember: it is not the Obama government, but the constitutional law, lawyers and law-makers who feel their obligation to the American tradition of liberty.

10

Meta-Power Game of Politics: Metamorphosis of the Nation and International Relations

The argument of this book is that the metamorphosis of the world is 'happening'. But what does 'happening' mean? In this chapter I outline the answer: metamorphosis, in socio-logical terms, is not a fate, nor is it something which follows from the law of nature, as in biology. The differences are, first, that we do not know the ending. Second, it is a politics of side effects involved in a power conflict between those who defend the national order and orthodoxy of politics and those who challenge it by rewriting the rules of power and politics. The cosmopolitan middle-range concept I am intro-ducing here is *the meta-power game of politics*.

By 'the meta-power game of politics' I mean that national politics, which worked in a rule-following way, and the new cosmopolitan world politics, which works in a rule-altering way, are fully intermeshed with one another. They cannot be separated in terms of specific actors, strategies or alli-ances. It becomes clear that, in the twilight zone between the passing of the national era and the emergence of the cosmopolitan era, political action and power are following two completely different and yet mutually interwoven scripts. There are two different players on the world stage, perform-ing two different plays in accordance with each perspective, such that there is a highly paradoxical interweaving between the established and the alternative political drama, between that which is defending the national world order of politics and that which is trying to change the rules and roles of the power game in a cosmopolitan way.

The analogy of the game needs to be interpreted carefully. The spaces of action do *not* operate like a game, wherein players adopt strategies in a competition with others to win, watched over by a referee. There is not one single game for all. Players play different games at the same time. In fact, cosmopolitization is defined by the turbulences arising from this fact. There are no rules of art, no *raison d'être* for the cosmopolitized spaces of action and no referee. Given that there are different rules (e.g., between boxing and rugby), it is no longer easy to identify the appropriate moves and to agree on what 'victory' and 'defeat' mean.

At the same time, the new, open-ended meta-power game cannot be played alone, much less according to the rules of the old nation-state game. The old game, for which there are many different names, such as 'nation-state', 'Westphalian order of sovereign states', 'national capitalism' or even 'national welfare state', is challenged because the metamorphosis of the world has introduced new spaces and frameworks for acting. Politics is no longer subject to the same boundaries as before or tied solely to state actors and institutions. Yet, it is possible that old and new actors are personified in one individual, who has to define and create their sub-political, sub-revolutionary roles and resources on the playing board.

In the metamorphosis from one era to another, politics is entering a peculiar twilight zone, the twilight zone of *double contingency*: nothing remains fixed, neither the old basic institutions and systems of rules nor the specific organized forms and roles of the actors; instead, they are disrupted, reformulated and renegotiated in a conflict between those actors or organizations which defend and those which try to change the national order of politics. Importantly, this metamorphosis of power politics is not simply about changing perceptions but about a real confusion of categories, scripts, plays, players, roles, doctrines, and spaces of action.

This conflict of the 'negotiation of metamorphosis' can be observed from different perspectives, from the perspective of globalizing capital or from that of actors in civil society movements. Here I would like to observe the change from the perspective of national politics, and specifically in relation to two case studies: the metamorphosis of the European Union and China's involvement in the dynamic of the global climate risk to humanity.

1   The Metamorphosis of European Politics

The EU is a prime example of the meta-power game. Europe is not a fixed condition, not a territorial unit, not a state, not a nation. In fact, there is no 'Europe'; there is the metamorphosis of Europeanization, a process of ongoing transformation. In the case of the EU, metamorphosis is another word for variable geometry, variable national interest, variable internal/external relations, variable borders, variable democracy, variable statehood, variable law and variable identity. One of the puzzles of political theory is the question of how nation-states develop cooperation within the context of state sovereignty without losing their identity and finding answers to global challenges.

The metamorphosis of nation-states into European forms of governance and cooperation is the big historical experiment to achieve this. The first step in this metamorphosis was the 'politics of side effects'. Although the process of Europeanization – the 'realization of an ever closer union of peoples of Europe', as the EU treaty puts it – was intended, its institutional and material consequences were unintended. The striking fact is that the process of integration did *not* follow any master plan. The opposite is the case: the goal was deliberately left open. Europeanization operates in a specific mode of institutionalized improvisation.

This 'politics of side effects' seemed for a long time to have one major advantage: even while the juggernaut of

Europeanization pressed ahead relentlessly, it did not appear to require an independent political programme, a fixed goal or a political legitimation. In the first stage, the metamorphosis of nation-state politics into EU politics could occur through transnational cooperation of *elites* with their own criteria of rationality, largely independent of national publics, interests and political convictions. This understanding of 'technocratic governance' stands in an inverse relation to the political dimension. The framework of European treaties exercises in this way a meta-power politics that alters the rules of the power game of national politics through the back door of side effects.

The 'invention' of Europe was a product not of public deliberation and democratic procedures but of judicial prescriptions and practice. It was and is the European Court of Justice (ECJ) which elevated the European founding treaties to the status of a 'Constitutional Charter' in key court decisions in 1963 and 1964.

Here we have another stage of metamorphosis – a kind of 'cosmopolitical takeover' – a process which was driven forward by 'legal conversion' in cooperation and conflict with the various national supreme courts and, what is more, which was adopted by the national governments and parliaments as a basis for their further operations. This 'cosmopolitical turn' of the European Court of Justice gave rise to an authoritative form of constitutionalism in Europe *without* a formal constitution, based on a practice of law-making. Europe is the product of political praxis without political theory.

European metamorphosis of politics in this perspective is a politics of institutionalizing the cosmopolitan horizon in cooperation with the national horizon by *practising binding European law*. From this a meta-power conflict between the national actors and defenders of national constitutional law and the cosmopolitan actors of European law exists to this day. Here we can observe what the meta-power game of politics means in the context of 'politics of law'. On the one hand, the old national politics of law worked by applying the constitutional law; on the other hand, the new politics of law on the European level works by changing the politics of law. These two are now fully intermeshed and cannot be separated from each other: the game can no longer be played alone. What unfolds is that the power of the National Constitutional Court is slowly but surely shifted to the European Court, which pushes the National Constitutional Court in its practice into a fundamental conflict: on the one hand, it is supposed to judge on the basis of the national constitutional law; on the other, it has to anticipate the metamorphosis of the national to the European law system and thereby disempower itself.

But this metamorphosis is not a one-way street. The euro crisis lets us trace how national thinking in Europe received a fillip and liberal economists, as well as politicians of all stripes, redoubled their efforts to shift the European frame of reference towards Germany as a political frame of reference. The result was and is a conflict over sovereignty in the unfolding threat to the euro, because the return to the nation-state was and is thwarted by the monetary policy regime of the European Central Bank currently in effect. One can speak of a 'Draghi euro', by which is meant an unwritten monetary and fiscal emergency policy that has acquired enormous influence over the fiscal policies of the member states as well. What speaks for it, what legitimizes it, is that this emergency policy can point the way out of the crisis, the one leading towards more, not less, Europe. The return to the national framework is being undermined and overridden, therefore, by the metamorphosis of the 'division of sovereignty' at the expense of national and in favour of European sovereignty that is the result of the pressure of the risk to the euro.

In finance policy as well, the law of action has shifted from the control of the German Ministry of Finance towards the only actor fully capable of acting in the crisis, the European Central Bank. In this movement of metamorphosis, it ultimately comes down to the question: Who determines financial policy in the eurozone in this monetary policy state of exception? This is also in the final analysis the question that was referred by the German Federal Constitutional Court to the European Court of Justice. As the first step towards the phasing-out of national sovereignty over fiscal policy, this development led in Germany to the foundation of an anti-euro party, Alternative für Deutschland (Alternative for Germany), in which the 'German national economists' who are in thrall to methodological nationalism united to form a protest movement.

The legal dispute over the policy of the European Central Bank in the euro crisis shows how this can lead to confusing conflict–cooperation relationships. On the one hand, the German Constitutional Court declared the matter to be outside its jurisdiction and turned to the European Court of Justice. On the other hand, in doing so, it as it were put a gun to the head of the European Court of Justice. If the European Court of Justice blocks the political and legal decision of the German Constitutional Court to show the European Central Bank its limits, then the Constitutional Court will refuse to comply with it on this point. This is symbolic of an epochal conflict in the metamorphosis between national and European law. Characteristic here is a certain ambiguity of interests. The court wants to save not only the euro but, above all, itself – from irrelevance in a European context that is increasingly important. In other words, the German Constitutional Court wants to consolidate its role and power in the European context as well. That is, it is conducting politics on its own behalf.

European metamorphosis does *not* mean that the nation-states disappear, but it does mean a 'Copernican turn': Europe is no longer turning around the nation-state as the sun seems to be turning around the earth; the nation-states are going to turn around Europe, just like the earth is turning around the sun. This means that the nation-state, or, even more, the idea of the nation-state, is metamorphosing.

But haven't the European elections in May 2014 and the success of the anti-Europe parties shown that cosmopolitan Europe is in decline, overthrown by the anti-Europeans?

What appears on first sight as a clear case is actually a fallacy of the national outlook. It misses the logic of the really existing metamorphosis of the EU.

The next step was that, for the first time in the history of the European parliamentary elections, the various party groups nominated candidates for president of the Commission. As a result, although the anti-European parties entered the European Parliament significantly strengthened, the president of the Commission, now elected in a European-wide ballot, was simultaneously legitimized and strengthened through a democratic election. The interplay and conflict between national and European politics in turn exhibits a confusion of responsibilities and positions of power. Thus, according to the rules in force, the election of the president of the Commission is made on the recommendation of the national heads of state and heads of government who share power in the European Council, but at the same time through the European Parliament. Following the election, this ambiguous constellation in the power triangle comprising the Commission, the Council of the heads of state and government, and the European Parliament unexpectedly shifted towards more Europe and more democracy. The national heads of government at least in part rejected the person of Jean-Claude Juncker as president of the Commission as nominated by the conservatives. The result was an epochal conflict between two conceptions of democracy. The one draws its sustenance from national democracy, which tries to stand its ground against European claims to power; that was the position of the British prime minister, David Cameron. The other position attached central importance to the fact that the European election of the candidate had lent the latter and the European Parliament new legitimacy and power. If this election were now to be boycotted by the European Council and undermined by the proposal of another, non-democratically elected candidate on its own authority, it would amount to a kind of assassination of European democracy. In this conflict, the German chancellor, Angela Merkel, in the end (contrary to her own dominant role in Europe) came down in favour of enhancing European democracy, and hence of the simultaneous loss of power of the European Council of the national heads of government. Thus, the elected president of the European Union – Juncker – became the actual president of the European Union.

This marked the completion of another step in the European metamorphosis. The Council of the heads of government, thus the nation-state power option in Europe, experienced a reduction in power, and European actors – the president of the Commission and the European Parliament – an increase. The new president of the Commission is trying to consolidate this shift in power, on the one hand, and, on the other, to give it a creative turn by forging an alliance between the European Parliament and the 'ruling' Commission. Under the treaty, the latter has the exclusive right to propose legislation in the EU. At the same time, if a clear ('qualified') majority of the European Parliament supports it in the first and second readings, then the Commission sets the direction, which is to say that the European Council no longer has the final say. In this way, the new president of the Commission, Juncker, is trying to transform its democratic legitimacy into a kind of European governmental power. The Commission makes the proposals, and the president of the European Parliament, Martin Schulz, is supposed to organize the necessary majorities – this is the new Commission–Parliament power axis that seeks to override the power of the national regional heads of government in Europe.

It would be completely mistaken to locate this metamorphosis of the European power structure exclusively in the relations of power between national and European actors and institutions.

As for the anti-European parties, they need to be understood as running for seats in the European Parliament in order to break the national away from Europe. Yet, if they had no seats in the European Parliament, nobody would care about their anti-European sentiments and goals. Here we have the paradox that the European Parliament is empowering a kind of politics that want to destroy European democracy.

2   How Climate Change Risk is Used to Renegotiate China's National Self-Definition

The power of metamorphosis and the metamorphosis of power is indicated by the *metamorphosis of opposition*. This concept refers to the power struggles between different forms of ignorance and rejection of bads and the staging of bads in the context of redefining nationalism with the sense of a cosmopolitan outlook. In the face of global risk, new normative horizons emerge which challenge existing institutions, especially national ones. The nation-state is in a dilemma. On the one side, it needs to align itself with the emerging new normative expectations; on the other, it is unable to cope with the global nature of risks, such as climate change. This chapter deals with this issue in the case of China, more specifically how the biggest newspaper, the *People's Daily*, reports on climate change. What we see in this example is how the risk of climate change, as a *global* issue, gets translated into a 'nationalized-cosmopolitan' issue, which, in turn, leads to a cosmopolitan turn in Chinese national politics and identity.

The promulgation of world risk society – this is my thesis – resolves in a reimagination of nationhood which takes place in the context of the staging and perception of climate risks to humanity, but also in the context of global financial risks threatening all sub-systems of societies or the human rights issues forcing nations to renegotiate their self-understanding in relation to other nations. The same is true with migration flows (e.g., direct encounters with difference), global terrorism (e.g., posing existential threats to civil societies and shifting the parameters of the Hobbesian contract), global generations (who have grown up digital and are struggling to become 'digital citizens') and global and local interpenetrations of world religions (e.g., the proliferation of ethno-religious diasporas), to name but a few of the global backdrops shaping the national and worldwide conflicts about redefining nationhood in the context of world risk society (Beck and Levy 2013).

China is a particularly interesting case study for two reasons. First, China, being the most important developing country, demonstrates that the normative horizon of climate politics has already been distributed around the globe. Second, in China the state authority plays a powerful role, especially in the *People's Daily*, which is a platform and public voice for the Communist Party.

The following pages examine the political and ideological discourse strategies embedded in the national framing of climate change issues. Drawing on Zhifei Mao's study 'Cosmopolitanism and the Media Construction of Risk' (2014a), it does this in terms of concerns about responsibility, consequences, conflict, morality, human interest and leadership.

There are two phases in the metamorphosis of the opposition of Chinese national politics which are apparent in the reporting of the *People's Daily*.

Phase 1: Climate change risk and the context before and during the Cultural Revolution

In this phase, the reporting of climate change in the *People's Daily* went through three stages: overlooking the issue at first, followed by the rejection of the issue, and then several years of ignorance. The 'overlooking stage' took place before the publication in 1973 of the first news article talking about climate change in relation to global risk, in the sense that climate change was not conceptualized as global risk in China because of what could be called 'unintentional silence'. The silence was unintentional in the sense that it was simply due to people's not knowing about the issue. Thus, for example, most of the Chinese people at that time would not conceptualize the extreme weather they faced with reference to the global risk of climate change.

Then, with the publication of the news item that denied that climate change was a global risk (Zhang and Zhu 1973), followed the phase of 'rejections of bads'. The interesting point is that the first time the *People's Daily* mentioned the term, it was referring to a large-scale shift in global weather patterns (during China's Cultural Revolution).

Before and during the Cultural Revolution the problem of climate change was relativized or ignored in the socialist belief of progress. As such it was subject to the strategy of categorically rejecting any idea of bads and risks as cultural pessimism.

The article entitled 'A Discussion of Climate Change in Recent Years', talking about the global 'abnormal weather' in 1972 (Zhang and Zhu 1973), was written by two meteorologists. It employed the consequence frame to represent the issue and denied the destructive consequences of climate change by declaring 'humans must conquer nature because socialism is anti-disaster'. The article started using technical language to describe the global concern about the unusual weather and introduced the average temperature movement as the method to measure climate change. This language was consistent with the technical words and non-political tones in media reporting of meteorological research before the Cultural Revolution (Xinhua News Agency 1962). Although the authors denied that climate change was a global risk by refuting the 'concerns and fears' of some meteorologists that humans could once more face an ice age or a small ice age, they did claim that the abnormality in weather was closely related to agriculture and the livelihood of people all over the world. At the end of the article, the authors shifted to the use of very political and ideological language in mentioning how the Chinese people conquered the drought disaster in 1972 and achieved a 'bumper harvest' under the leadership of Chairman Mao, praising Mao's strategy and the good mechanism of socialism.

The article seems to adopt the discourse of 'silence detoxifies' (Beck 2009: 193): the expression of risk is silenced or marginalized to ensure the smooth running of the current social and political system. However, its publication had other meanings. Here the specifics of the discourse about climate change in China are important: it is not so much about constant protest and activism from below (as was the case in some European countries) – that is to say, the 'silence detoxifies' discourse in some European countries occurred *after* the staging of climate change by green groups and some scientists. However, in the case of China, there is a paradoxical moment: this discourse – against the political will behind it – served as the implicit acknowledgement of the urgency of climate change issues. The news article, though denying that climate change was a global risk, broke the silence of the overlooking stage and showed millions of Chinese readers that there was an alternative explanation for the unusual climate in their daily life: climate change on the global level.

Thus, after the 1973 article followed a stage of *intentional ignorance*, which is the continuous 'silence detoxifies' strategy instead of the previous unintentional overlook of the issue. The difference between the two periods of silence lies in the knowing or not knowing, the possible association between climate change and the concept of global risk. When the issue was raised in 1973 there was no flashback. The topics of climate change, globally, and risk have remained 'planted' in the Chinese people's frame of reference ever since. This way, in a first step, the beginning of a new world horizon was created and affirmed. At least in this case, the counter-discourse against the statement that climate change is a global risk does not serve only as 'silence detoxifies'.

Phase 2: After the Cultural Revolution

In the second phase the metamorphosis of Chinese national politics and identity becomes apparent.

After the Cultural Revolution, climate change appeared in the title of three news items in the *People's Daily* between 1977 and 1987. Compared with the articles published during the Cultural Revolution, there were three major changes in the framing of climate change: first, the news frames of these articles became diverse. While the journalists were still concerned about the consequences of climate change, such as the world's energy problems, population problems and food problems (Zheng 1979), they did start to highlight other aspects, such as human interests under the issue of climate change and the attribution of responsibility for causing climate change (Xinhua News Agency 1980). Second, unlike the article published in 1973, one item published after the Cultural Revolution partially admitted the negative consequences of climate change on people by stating that 'some of the areas will benefit from climate change, and some will be hurt' (Zheng 1979). It also mentioned the First World Climate Conference and highlighted the issue of carbon dioxide emissions. Instead of declaring 'humans must conquer nature', this author showed his faith in science to 'know the patterns of climate change'. Lastly, the ideological terms of socialism and Chairman Mao were not mentioned in the relevant articles during this period. All three articles frequently employed technical language in framing climate change.

The redefinition of national responsibility is manifest in the way the central term 'we' was used by Chinese representative Luo Xu. He claimed that climate was crucially important to the world and that 'there would be serious consequences if *we* do not take actions' . He used an 'inclusive we', referring to China and other nations in order to share the global responsibility of dealing with the issue. This was nationalized through another news article:

Facts speak more than words. 2009 is the most difficult year for China in economics since the 2000s. However, for the protection of climate and the environment on earth, Chinese people have been and still are acting as actively and seriously as they could, and show a great image of a responsible country to the people all over the world. (Lin and Yang 2009)

However, the 'inclusive we' was broken up again and replaced by the distinction between the developed and the developing world, between the capitalist North and China. Since the very early years of China's attendance at discussions on climate change on a global level, the *People's Daily* used the responsibility frame to highlight that the developed countries had caused the problem and thus needed to take more responsibility, while the developing countries, including China, should not be pressurized (e.g., Xie 1991; Xinhua News Agency 1994; Zou 2007).

Twenty-one of these articles reported how the Chinese premier, Wen Jiabao, effectively and efficiently engaged in dealing with the issue of climate change, both nationally and internationally. These articles emphasized how Wen protected the national interest by insisting on China's own strategy in reducing greenhouse gases while evoking a sense of cosmopolitanism by taking responsibility to solve the global problem. A typical news item published on the front page of the *People's Daily* reported Wen's attendance at the 2009 United Nations Climate Change Conference as a productive discussion with political leaders from both the developing and developed countries, as well as his 'sincerity, confidence, and being strong-minded' (Zhao et al. 2009). In this sense, climate change was employed as a platform for polishing the image of the political leader.

Hence, what we see is that the metamorphosis of opposition occurs in the coming together of four components. First there was the staging of a cosmopolitan sensibility, which took seriously the global character of climate change. Second, playing into this, there was a redefinition of the national self-understanding as open and responsible. Third, China positioned itself as a Third World country against the dominant North. And, fourth, there was the personal striving of the then leader of the Communist Party to present himself as a responsible and open-minded leader.

What is interesting in the Chinese case is that metamorphosis is not about protest and activism from below but is envisioned, initiated and driven forward by those in power and used by them to bring about a change in the leadership within the Communist Party.

11

Cosmopolitan Communities of Risk: From United Nations to United Cities*

From a national outlook, cities play an interesting but not a distinct role in world politics. They remain sub-actors within the national and international frame. Looking at the metamorphosis of the world from a cosmopolitan perspective, the relationship between states and cities is reversed. In the face of global and cosmopolitan risks, states remain locked in the fiction of egoistic sovereignty and fail. Cities, however, are not locked into the fiction of the national container. On the contrary, historically they often held an autonomous position. Faced with global risks, they are more open to cooperative cosmopolitan politics. As a consequence, the relationship between states and cities is transposed. Cities turn into pioneers which take up the challenge of cosmopolitan modernity as an experiment to find answers to the world at risk. Hence, the framing of cities as cosmopolitan actors sheds light on the metamorphosis of international relations as well as international law-making.

This chapter is not about the role of single cities in world politics but about the cooperation in the face of global risks between a number of them, which play an active role in the

---

* This chapter was written in collaboration with Anders Blok.

cosmopolitan spaces of action. In this respect, cities are specific actors and different from other (sub-)political actors, such as civil society actors, market players, and religious movements and organizations. Cities can take collectively binding decisions. They are embedded in an active way in international law-making. And they are subject to traditional democratic practices and challenges; their mayors have to be re-elected based on how much they have achieved in the face of these challenges.

But how do cities turn into 'imagined cosmopolitan risk communities'? This chapter introduces the concept 'imagined cosmopolitan risk communities' as a middle-range concept for cosmopolitan theorizing. To be clear, the notion of 'community' differs from the notion of 'network'. Community is not only about being connected and interdependent; it is also more than exchanging information and regular meetings to discuss shared problems. Indicators of the characteristics of risk communities include projects of combined legislation and political decision-making and forms of civic participation beyond the borders of cities. But these are projects in process. They could be manifest in institutions but, at present, are more likely to be 'just' apparent in unfolding projects. The situation is currently shaped by 'municipal voluntarism' (Bulkeley 2013).

The sociological observation of these processes could be part of the public discussion and emergence of these institutions.

1   The Metamorphosis of World Affairs as Seen Through the Lens of World Cities

In order to elaborate on these theses, it is useful to return to the distinction between transformation and metamorphosis. Looking at political decision-making and collective action through the frame of reference of transformation means focusing on the problematic of national politics (e.g., elections, changes in the constellation of political parties, changes in the national and regional order, etc.), as well as

on international organizations, alliances, regional wars,
'failed states', etc. From this perspective, world cities appear
to be of minor political importance with respect to the new
challenges we are facing today.

Within this national framework, the priorities are clear.
Everything revolves centrally around a geopolitical shift in
power, where the reproduction of the national-international
order of world politics is always tacitly assumed. At present,
much attention is being devoted, for example, to the question
of whether, in a couple of years, China will have displaced
the United States from its position of global dominance;
whether the Arab states will sink into chaos or be overrun
by militant fundamentalists; or whether the European Union,
which is not able to speak with one voice, is being marginalized in spite of its global economic position. Here the anar-
chic character of world politics that is posited as a constant
in the model of international politics is not applauded but is
instead defended with an argument *à la baisse* – namely, that
it is the best among the worst systems. Any other system
seems inconceivable to the national outlook or appears to
lead to chaos (so it is argued).

Metamorphosis here means the opposite – that national
and international politics are seen within the framework and
through the lens of world cities and their emerging power in
world affairs. This change in the frame of reference first
opens our eyes to the actual metamorphosis of the world
that is occurring in the interdependence and competition for
power between nation-states and world cities and opens up
new perspectives on cosmopolitan climate policy.

- The emancipatory potential of climate change risk is
  apparent not within the reference horizon of nation-
  states but within that of the world cities. The United
  World Cities, not the United Nations, could become
  the cosmopolitan agency of the future because, by
  comparison with nation-states, alliances of world
  cities are acquiring a new sovereignty, power and
  pioneering role in world politics, which is confronted,

on the one hand, with global risks and, on the other, with the fact that nation-states are more or less capitulating in the face of these challenges.

- A different political logic becomes apparent, one which switches from the national friend–enemy logic to the cosmopolitical logic of cooperation, which – it must not be forgotten – also includes an existential conflict logic. This has epistemological implications. It expresses the essence of what is meant here by the 'metamorphosis of the political'.
- In this way it once again becomes clear why we must replace methodological nationalism with methodological cosmopolitanism: methodological nationalism, the national frame of reference, blinds us to the rapid metamorphosis of world politics and, hence, to questions that can arise and be analysed only within the cosmopolitan outlook, which accords central importance to the new global political role of world cities.

2   Cosmopolitan Communities of Risk

Who and where is the actor or bearer of cosmopolitanism in the age of global risks? In answering this question, two mistakes can be observed: the first one is pessimism – there are *no* actors or bearers of cosmopolitanism. This is the 'realistic' answer of those whose thinking is locked in the categories of the nation-state.

The opposing view argues that we are witnessing the birth of a new world revolutionary subject, the Union of World Cities. The problem with this position is that it repeats the error of socialism but in a different way: world cities take the place of the working class.

In contrast, I propose a third position, which is primarily of an empirical-analytical nature. According to this position, world city politics is transformed into translocal world politics, connecting local and global governance – in competition and cooperation with national-international world politics and in cooperation with the global sub-politics of civil society movements. This third perspective accords central importance to the metamorphosis of urban political space. To this end, I introduce the notion of *cosmopolitan community of risk*. This concept, which is key for cosmopolitan theorizing and research, combines the following constitutive conditions.

1   *Communities of global risk*: In the experiential space of world cities, invisible risks often become visible. Just think of the smog that engulfs world cities in poisonous clouds. In this way, world cities become the reflection and symbol of the emancipatory catastrophe. The world of nation-states stands for failure because, in their national egoisms, they block each other. The world cities, on the other hand, stand for the interplay of collapse and awakening. Here the clash of global risks becomes a matter of everyday experience, but also the clash of global inequalities, the clash of world conflicts (the Middle East conflict is played out on the streets of Paris, London, Berlin, Rome, etc.) and the battles between suicidal and survival capitalisms. Here the clash of the global diversity of different groups, with their different ways of seeing the world, being in the world, and imagining and doing politics, is also an everyday occurrence. The world cities are in this sense a field of experimentation of cosmopolitanism: How can the differences in the world and world history coexist in a single political place?

However, the painful and conflict-ridden, everyday experience of the problems created by global risks in the milieu of world cities is a necessary, but not sufficient, condition for the formation of a community of world cities that pursues shared goals, such as the implementation of an efficient climate policy. Those who, like many climate scientists, conclude from the insight into the imminence of a climatic apocalypse that a metamorphosis of politics follows from this as a matter of 'rational' necessity, succumb to a fallacy. Many, especially highly committed, climate scientists have manoeuvred themselves into this cul-de-sac. They are utterly unable to comprehend why people of all nations, religions and ethnic groups, rich and poor, men and women, blacks and whites, do not finally turn into *Homo oecologicus* out of their most essential interest in survival – why all human beings do not become little climate scientists. This inference from the 'objectivity' of the recognized problem to the metamorphosis of political action and everyday behaviour is sociologically and politically naive. When they notice that people fail to behave like little climate scientists, they feel compelled to downplay democracy, indeed to regard it as antiquated, and to seek the solution either in 'Gaia' or in variants of environmental dictatorship.

2   *Cosmopolitan communities*: The constitutive condition of the everyday experience of global risks must be supplemented by the question of whether and how a learning process from below is looming that makes it possible to give the conflicts of explosive diversity a constructive political turn. What is at stake here is whether a cosmopolitan common sense is taking shape in the milieux of world cities. In other words, the first condition points to the 'cosmopolitan *community of risk*'. The second condition, by contrast, raises the question of the origin of '*cosmopolitan communities* of risk'. Can the hope that springs from the everyday, often brutally anticipated decline acquire political power?

The milieu of world cities is in this sense characterized by the ubiquitous, all-pervasive cosmopolitization in everyday life, including the irritability that incorporates an experience of generalized alienation of all against all, especially for the locals in their own, familiar world. Here a place of learning, a diktat to learn, is taking shape that no one can escape – a 'cosmopolitanism for the immobile', a cosmopolitanism that springs from the resistance against the growing hostility towards foreigners.

If one accords central importance to world city climate policy, then it is also important to pose uncomfortable questions, namely: How do the politics of climate change necessitated and mobilized by world cities deal with the fact that, in other cultures and parts of the world, whose inhabitants likewise also live in world cities, it is not even climate deniers who set the tone and the term 'climate change' is non-existent? What connections can be made between climate change, explosive diversity and metamorphosis of the world? Is world city climate policy being thwarted by global imbal-ances that are also at the source of the climate risk within and between the world cities of the South and the North? Or could, on the contrary, the migrants who cultivate 'spa-tially polygamous' ways of life and thinking become climate advocates in their families or countries of origin or within their own networks?

The notion of risk communities presupposes the concept of concern or 'care'. With the anticipation of the catastrophe, concern for oneself becomes concern for everyone else. Thought through, this means two things: concern for oneself includes concern for one's enemy; but it also means that, out of the concern for everyone and everything, new enmities, new existential conflicts, arise – conflicts over those who violate this concern.

If one understands the communities of world cities in this sense as '*cosmopolitan communities* of global risk', however, one must abandon the widespread assumption in the social sciences that community-building is possible only on the basis of *positive* integration through shared values and norms. Instead, it supports the thesis that another form of community-building is also possible, one which arises in the course of conflicts over *negative* values (crises, risks, threats of annihilation) – the thesis of emancipatory catastrophism.

A further step is necessary: the solution is not a product of technocratically enforced, top-down consensus. Given the manifest nature of the problems, can a bottom-up conflict discourse that is both urban and addresses a global public arise in the milieu of world cities, a discourse from which the claim to networked, communal action proceeds? Of course, this is also in the first place only a question to be explored.

Therefore, we do not assume that the global character of the apocalyptic risks gives rise in itself to the commonality of political action. Everyday experiential space of cosmopolitan interdependence does not arise as a love affair of all with all. It consists in and arises out of the universal indignation over the everyday plight of global risks that have become irrevocably apparent. In this way the global urban imperative can unfold its power – cooperate or fail! This is what we mean by urban 'cosmopolitan realpolitik' (Beck 2005). In this way, a definition of threats that is accepted by people in their everyday lives across all national, ethnic and religious boundaries and divisions can take shape. As a result – in accordance with the thesis of emancipatory catastrophism – a common space of responsibility and action can be created which, analogously to and in competition with the national space, *can* (but by no means must) found democratic political action. This is the case when the accepted definition of threats, on the one hand, is made brutally visible in the everyday life of world cities but, at the same time, also becomes answerable within the various horizons of individual action. On the other hand, it also leads to global norms and agreements because of the global political role of world cities. In this way, world cities, in particular – unlike nation-states – present the opportunity of overcoming the well-demonstrated difficulty of moving from the shared definition of threats to binding practical commitments. In any case, the cosmopolitan, democratic, community-building potential for indignation inherent in climate change is becoming enormous and evident, especially in the milieu of world cities. Nowhere other than in world cities and their informal and formal connections is the opportunity to shape the potential for indignation, the power of the anticipated catastrophe, into institutional, democratic political forms so palpable.

3 *Sovereignty and formative power of world cities*: As we have just seen, the concept of 'cosmopolitan communities of risk' involves both (1) 'communities of *global risk*' and

(2) *'cosmopolitan communities* of global risk', as well as the connection between the two. In this sense, world cities stand for the interplay of collapse and awakening. But that's not enough. To this must be added, as further key conditions, *world city sovereignty* and *political and legal organizational power* on the local and the global level. The question of how far world cities become detached from the nation-state space of government and jurisdiction and form cosmopolitan communities may – obviously – not be answered in the same way for all world cities. Empirical studies show that this is more true for American and Western European world cities, but less so for Chinese and Russian world cities.

Here, the question of the sovereignty of world cities should not be equated with the question of the extent to which world cities can or cannot liberate themselves legally and politically from the dominance of their nation-states. There is also such a thing as 'urban global domestic politics'. For instance, cities have some autonomy in questions of inner-urban settlement of migrants. They can decide whether to follow a separation strategy that locates different national and religious groups in different neighbourhoods or to follow a model of multiculturalism – that is, a strategy of mixed settlement of migrant groups. Then the concept of 'United' Cities would also acquire an inner-urban meaning.

At the same time, studies also indicate that the links between world cities, and their participation, and indeed their initiating role, in global norm-building processes are on the increase. In this way, an embryonic political structure of United Cities has become visible. In their database, Bulkeley et al. (2012) list sixty different transnational initiatives that have sprung up during the past decades, among them the C40 Cities Climate Leadership Group, the Covenant of Mayors, the Cities for Climate Protection (CCP) programme and, not least, the umbrella organization United Cities and Local Governments (UCLG), which was established in 2004.

Finally, it is surprisingly clear that the political-ideological orientations that bear the predicate 'conservative' are losing their ability to win majority support in the explosive experimental milieu of world city diversity. A global political map depicting the relationship between national and world city politics should make clear that world cities stand out as colourful islands in the black ocean of national-conservative politics. This applies to New York, as well as in different ways to London and Seoul, and even to Zurich.

It is ultimately the professional intelligentsia working and living in transnational networks, combining competence and (environmental) experimental spirit with economic success, whose voice is becoming more influential in the competition for power in the world cities. To what extent one can already speak of a class of professionalized intelligentsia as a source of ideas and a bearer of power remains an open question, both politically and sociologically. Yet, world cities are certainly home to large populations of young middle-class professionals who are growing increasingly disenchanted with capitalism as usual, and who want to explore and implement new green alternatives. To use the language of social movement theory, world cities are where we find the various emerging 'reform nexuses' of capitalism (Chiapello 2013). These nexuses bring actors who share certain green critiques of capitalism into new political constellations, including activists, consultants, labour unions, green-tech businesses, green policy-makers, and so on. The greening of capitalism starts in cities.

However, the fact that conservative party ideologies and conservative candidates who are still trumps in the competition for power at the national level have lost their majority appeal in world cities speaks this language. An additional factor is the majority appeal of transnational world city professionals. Taken together, this can be understood as an indicator that something like 'cosmopolitan communities of global risk' are in fact taking shape. Then the 'axis of the world cities' would constitute a trace of the future in the present. Translated into political terms, this is captured by the formula 'from United Nations to United Cities'.

The unfolding process of metamorphosis of communities of world cities can be explored further through three different case studies: the metamorphosis of traffic, the expropriation through risk, and the metamorphosis of conflict.

3 Metamorphosis of Traffic

The metamorphosis of traffic is obvious. Things that not long ago seemed outdated, such as the bicycle, are back, in the sense that they have come to be valued again. In turn, what used to stand for progress and prestige, the car, has come to be devalued as a source of risks and bads. As the culture of automobility has spread worldwide, it has come to shape much of urban life and spatial form – for instance, in subordinating other 'public' modalities of walking, cycling, travelling by rail, and so on, to the 'quasi-private' mobility spaces of ubiquitous car traffic. In recent years, however, under the weight of growing ecological and health concerns, urban commitments to car driving are undergoing serious re-evaluation all over the world. With cars increasingly perceived as cumbersome and 'dirty', cities are experimenting with a range of alternatives, from expanding public transport and bicycle lanes to new forms of mixed and compact urban development that reduces the need for mobility. In this alternative approach to sustainable urban planning, pedestrians and cyclists are at the top of the new 'transport hierarchy', while cars are at the bottom (Banister 2008).

In recent years, certain cities have come to be known for their efforts at 'greening' transportation policies, and this knowledge travels and is mobilized also by other cities. For instance, following a protracted political struggle in the early 2000s, London is nowadays recognized as a city that has effectively implemented congestion charging to reduce inner-city traffic. And, when it introduced its congestion charging system in 2006, Stockholm could build on the example of London. Similarly, the South Korean city of Changwon, having identified Paris as a city that had implemented a successful bike policy to reduce carbon emissions, sought to learn from its public-sponsored 'Velib' bicycle programme in introducing its own (Lee and van de Meene 2012). The city of Yokohama has teamed up with Nissan to run an electric vehicle car-sharing system and has plans to expand the system throughout urban Japan (and possibly beyond). Meanwhile, downtown Manhattan is undergoing 'Copenhagenization' in an effort to learn from the Danish city's extensive historical experience of how to increase the use of bicycles and reduce the use of cars.

Common to all the cases is the fact that urban planning and transportation practices and policies are changing worldwide, not via some global exercise of statutory legal power but, rather, through the power of the 'good example' (or 'best practice'). Moreover, all of the cases involve contestation and conflict on the ground. Urban mobility (and other) infrastructures are obdurate; they cannot simply be changed overnight. Fears are expressed by some that reducing car traffic will dampen economic activity and growth. Yet, while practical implementation remains piecemeal, urban *norms* and *visions* have been transformed entirely, subject to the metamorphosing power of global risks: who, nowadays, would ever dream of promoting fossil fuel-based cars as the future of urban mobility? Whatever the difficulties involved, city streets have become experimental sites for new norms of urban living.

What we are witnessing in the space of urban climate politics is a *transnational process of norm generation*, radically changing what counts as innovative, visionary, and legitimate urban planning and development. In this process, a growing number of inter-city network alliances play key roles in generating, sharing and helping to implement new policy-relevant know-how on urban greening across geographically distant cities worldwide.

Together, existing city alliances form a complex organizational architecture of overlapping, 'networked' and transnational spheres of urban authority, which is changing the entire landscape of global climate governance. In part spurred by the lobbying of the C40 and other urban alliances, cities are increasingly gaining recognition and voice in international law, the United Nations, and other global governance forums previously restricted to nation-states. All of this rests on the realization, on the part of municipal governments, that sharing and pooling their authority across transnational boundaries is the only way to start addressing the common challenges posed by the global risks of climate change.

On the one hand, we should not underestimate the power of normative pressure that 'urban peers' can exercise: as we said, urban visions and norms are indeed changing, in the direction of greening, sustainability, and low-carbon transition. Together, cities are creating, and being enrolled in, new global moral geographies of carbon, creating new shared norms of what it means to exercise responsible and accountable urban development.

On the other hand, sooner or later, we will need to raise the hard questions of collective decision-making and pooled sovereignty at the urban level: How exactly are we to link 'the urban ground' to new globalized regimes of legally and politically binding authority (see Sassen 2012)? What, indeed, would an overarching architecture of United Cities look like, if not simply modelled on the (often ineffective) United Nations?

4 Expropriation Through Risk

As I have argued elsewhere, risk is not catastrophe but the anticipation of catastrophe. Hence, one of the key moments of metamorphosis is that the very *anticipation* of catastrophe devalues capital.

The very anticipation of urban flood catastrophes carries vast socio-political consequences – consequences which are, however, often lost in the blind spots of 'technical' analysis. In the city of New York, for instance, current methods for estimating future flood hazards at city-wide level tend to 'undercount' more vulnerable sub-populations (based on income, race, etc.), leading to environmental justice concerns in terms of preparedness and relief effort (Maantay and Maroko 2009). Conversely, the impacts wrought by Hurricane Sandy on parts of lower Manhattan in October 2012 were as much symbolic as material: intensified flooding, this event showed, raises the prospect of turning attractive urban spaces for living and production into 'risk-spaces', thereby seriously devaluing them.

Small wonder if, as *Nature* reported, Hurricane Sandy 'sweeps US into climate-adaptation debate' (Tollefson 2012). Taken to its logical conclusion, the kind of 'ecological expropriation' exercised by major and more frequent flood events contradicts and violates the interests of the very institution of ownership in even the wealthiest pockets of modern, urban-based capitalism. Meanwhile, the distribution of urban flood risks also exacerbates already glaring global socio-material inequalities (Beck 2010, 2014): whereas large-scale engineering efforts are now under way to 'climate-proof' lower Manhattan, no such resources are available to vulnerable urban communities in the global South.

5 Metamorphosis of Conflict

In regard to the metamorphosis of conflict, two insights are crucial. On the one hand, global risks bridge friend–foe differences. On the other hand, new polarizations arise for which we are not yet sensitized enough and lack descriptive vocabulary.

The emerging urban climate alliances are part of highly fractured and unequal global relations, which give rise not only to new forms of cooperation but also to new kinds of competition, conflict and exclusion. To start with, the most powerful urban alliances, such as the C40, tend disproportionally to represent the wealthy large-scale cities of the global North – to the detriment of cities in the global South, clearly, but also excluding smaller and more 'ordinary' cities everywhere. Moreover, while the distinction between urban and rural areas is increasingly blurred – due to the global reach and spread of urban economic metabolisms (involving also energy, water, waste and so on) – this distinction is still highly politically relevant in all those contexts where land, resources, and living conditions are being redistributed (sometimes violently) via processes of rapid urbanization.

As a factor of increased environmental disruption, resource scarcity, and so on, the risks of climate change, and the way these risks are dealt with in cities, thus need to be situated within a wider framework of changing global inequalities that are multifaceted, ambivalent and open-ended (Beck 2010). And cities everywhere are at the very core of these new political alliances and cleavages, this changing political landscape of the twenty-first century that is being shaped by anticipations of global risks.

One dimension of such new cleavages has to do with the pursuit of 'strategic eco-urbanism' (Hodson and Marvin 2010) by local governments all around the world, particularly in wealthier urban contexts. Among other things, this involves targeted investments in 'eco-city' development, a set of policy practices that has emerged and spread globally since the early 2000s as part of inter-urban knowledge transfers, the rise of climate concern, and a clear urban policy preference for 'green-smart' technological solutions. Larger-scale eco-city initiatives have by now been launched in hundreds of cities worldwide, concentrated mostly in Europe and (East) Asia and with less frequency in South America, Africa and the Middle East (Joss et al. 2013), reflecting global urban inequalities. Most of the time, eco-city development is officially promoted as a new avenue for cities to attract investments, gain new markets, and brand themselves as 'global' and 'advanced' spaces.

This metamorphosis, as noted, is ambivalent and open-ended, with new forms of cooperation and competition intertwining and co-shaping one another into new economic and political landscapes. In some regions of the global South, for instance, appeals to the urgent need for climate adaptation provide one avenue through which local governments can now tap into international finance, thereby helping them improve urban infrastructures and life conditions for the poor. In other contexts, the pursuit of urban greening in the North has proven to generate its own unwanted side effects in the South. To expand the use of electric vehicles, for instance, requires the extraction of lithium from mines in Argentina, Chile and Bolivia, thereby getting caught up in a sensitive politics of land-rights, indigenous groups, and so on. Most of the time, these issues are excluded from view in urban centres in the North. However, from a cosmopolitan perspective, they should be made visible, and new institutional mechanisms should be invented for dealing with them in just ways.

Apart from these issues of global inequality and competition, and the extent to which they can be tempered through new forms of transnational urban solidarity, climate-induced risks may also be said to come with their own 'strategic' prerogatives. This is experienced every time a new storm surge or flooding event hits urban centres around the world, making the risks of climate change tangible and urgent. Such material urban realities may carry more weight than the abstract norms and future-oriented 'oughts' of global climate governance. When these realities hit cities, they hit hard, as more and more cities are learning. Alongside the mitigation of carbon emissions, adaptation and urban resilience are rightly becoming key priorities on urban agendas worldwide. In this context as well, framing adaptation as a matter of urban rights and justice is central to opening up its transformative potential.

6   The New Urban-Cosmopolitan 'Realpolitik'?

To summarize the argument so far, what I am suggesting is that urban politics, driven by concerns with global climate risks, is currently undergoing a fundamental metamorphosis and is manifesting itself in new urban alliances of transnational norm-generation, in new strategic investments for the making of eco-cities, and in new reform coalitions seeking to 'green' the functioning of global urbanized capitalism.

These tendencies and transformation involve all kinds of new ambiguities and conflicts. World cities, I suggest, are the main places where the clashes of global risks become matters of everyday experience and politics. When we speak of world cities as forming a cosmopolitan 'community' of global risks, then, this term does not stand in opposition to, but instead *includes*, such clashes and conflicts.

To this notion of cosmopolitan risk community, we now argue, corresponds the notion of a new and emerging urban-cosmopolitan 'realpolitik', a new pattern of alliance-making and conflicts that shapes urban politics worldwide (albeit in very different ways across different places and contexts). A matter neither of 'idealism' nor 'realism', this new realpolitik weaves together in new patterns what was previously thought of as separate: cooperation *and* competition; economy *and* environment; equality *and* inequality; solidarity *and* self-interest; localism *and* cosmopolitanism. None of these binaries works any longer if we want to capture and diagnose the metamorphosis of urban political decision-making.

Instead, we see new constellations of local and transnational actors, alienating former partners and allying strange bedfellows in the pursuit of a welter of interests and aspirations under the banner of umbrella terms such as 'sustainability' – itself a new meta-discourse of urban planning, embedding all sorts of value conflicts. In these new political constellations, new normative horizons of urban responsibility for low-carbon transition sit alongside, and are co-shaped with, new understandings of urban self-interest in a resource-scarce world. The resulting clashes, mobilizations and experimentations become tangible and salient in world cities in ways that are not the case in the 'abstract' political space of nation-states. This, more than anything, is why world city alliances are the new spaces of climatic hope: no other form of organization is better equipped to experiment with, invent and actually implement the new multi-sited architectures of political decision-making for the twenty-first century.

In order to realize this potential and work towards the vision of United Cities, however, political actors need to embrace, rather than shy away from, all the new ambiguities and conflicts of urban greening. If they do not, critics (e.g., Swyngedouw 2010) will be right in warning us against the post-political tendencies of 'sustainability', whereby urban climate initiatives are reduced to technocratic forms of infrastructural intervention, consistent with a neoliberal emphasis on the entrepreneurial city as a space of capital accumulation. Such a critique, however, is by no means a foregone conclusion. It is countered, in entirely practical and empirical terms, by the multitude of ways in which issues of public participation, environmental responsibility, carbon accountability, and transnational climate justice are *also* fully present on the urban-cosmopolitan real-political agenda.

What we need, more than anything, is a better grasp of how to navigate and analyse these new political landscapes. This is what metamorphosis is all about, and this includes the social sciences themselves: we need new ways of seeing the world, being in the world, and imagining and doing politics. What we propose in this chapter – in terms of risk community, urban-cosmopolitan realpolitik and the vision of United Cities – is meant as stepping-stones in this direction, increasing our abilities to see the changing world anew.

7   Outlook: A Reinvention of Democracy?

How much climate change can democracy endure? How much democracy does climate protection require? How is democracy possible in a time of climate change? Or, to put it even more bluntly: Why is the further development of democracy a *conditio sine qua non* for a world city cosmopolitan politics of climate change? These are extremely urgent questions. With the disastrous news of the rapid melting of the ice caps, there is a danger of succumbing to the fallacy of invoking a kind of emergency expertocracy that enforces the common good of the world against national egoisms and 'antiquated' democratic reservations in the universal interest in survival. Three components – the anticipation of the calamity engulfing humanity, the time constraints, and the progressively apparent inability of democracies to take decisive measures – mislead especially the most committed individuals into entertaining, at least half-unspokenly, Wolfgang Harich's vision of the 'strong, resolute allocation state' and of the 'ascetic distribution state', and hence models of environmental dictatorship (Harich 1975). The models of environmental dictatorship always take the hard, technocratically decisive individual state or world state as their starting point. But how are states supposed to impose the eco-consensus on other states, or how is the world state to impose it on other states? Through military threats, hence wars? This is not only a perspective that joins downfall to downfall but, at the same time – thank God – is plucked completely out of thin air, and hence is completely unrealistic. Here it becomes clear that the technocratic temptation rests on precisely the opposite of that to which it appeals: not on a sense of reality, but on the loss of the same.

The world city perspective, by contrast, shows that effective climate policy that exploits the emancipatory potentials of anticipated disasters is possible and real only from the clash between global diversity and global risks in an urban environment, hence only in the active participation of citizens, in the revival of democracy from below against the expertocracy. The world city is a site of experimentation of new forms of climate citizenship, new ways of inhabiting the world and new ways of reinventing democracy: first on an urban scale, then in the shape of polycentric and multi-scalar political alliances. Here, democracy means not just a set of procedures for political decision-making. At stake, more fundamentally, is what Clive Hamilton (2010) calls the 'democratization of survivability' in a world of intensified ecological threats and radical global inequalities.

Given the nation-states' resistance to cross-border collaboration and cosmopolitan politics, the 'turn to the city' is important, both epistemologically and politically, in order to discover or establish alternative institutions for cosmopolitan communities of shared risk, addressing the multiplying problems of a cosmopolitanized modernity without surrendering the democracy that nation-states have traditionally secured. 'In order to save ourselves from both anarchistic forms of globalization, such as war and terrorism, and monopolistic forms, such as multi-national co-operations, we need global democratic bodies that work, bodies capable of addressing the global challenges we confront in an ever more interdependent world' (Barber 2013: 4). Nations, inclined by their nature to rivalry and mutual exclusion, seem to be part of the problem and not of the solution in the world risk society of the twenty-first century.

In a metamorphosed world, global cities might reconquer a central position similar to that which they inhabited long ago in the pre-national world. Humankind began its adventurous path towards politics in the 'polis' – the city. The city was democracy's pioneer. But, for millennia, cities relied on monarchy and empire, and then on newly invented nation-states, to produce and reproduce social and political order. Today the nation-state is failing in the face of global risks. The cities – in history the social ground for civic liberation movements – might in today's cosmopolitized world of global threats once again become democracy's best hope.

Part III

Outlook

12

Global Risk Generations: United in Decline

This chapter focuses on the 'generation of metamorphosis' and the 'metamorphosis of generation'. The problematic of generation is a prime example in which the figures and moments of the metamorphosis of the world come together.

What does it mean to grow up in a 'divided' world – that is, in a world in which predominant exemplars and institutions (i.e., the dominant 'others' of socialization – teachers, politicians, judges, academics and intellectuals) convey and live according to a worldview shaped by a 'national outlook' while, at the same time, the metamorphosis of the world works inexorably towards the dissolution of the national world? How can one live with, indeed survive, the permanency of a metamorphosis when nobody can say where it is heading, a metamorphosis that affects the centre and the periphery, the rich and the poor, Muslims, Christians and seculars alike, a metamorphosis that does not arise from failure, crisis or poverty but that grows and accelerates with the successes of modernization, a metamorphosis that is not halted by non-action but speeded up by it? What does it mean for the political self-understanding of this generation, for their lifestyle, their consumer behaviour and their sense of hope and despair? Is the indifference of large parts of the younger generation the prerequisite for a pointed engagement or the sign of an unconditional surrender? How is one to behave if 'working' institutions fail?

What is specific about the understanding of the notion of generation in times of metamorphosis is that it has to be developed from *within a historical sociology of time* – i.e., a dynamic cosmopolitan sociology. In order to make this possible, I introduce the middle-range concept of *global risk generations*.

1 Metamorphosis of Socialization: the Disempowerment of the Older Generations and the Empowerment of the Younger Ones

Karl Mannheim, the founder of the sociology of generations, argued in 1928 that the concept of generation implies that their unity arises from their action. In this sense, generations are essentially political. Their transformative power is grounded in the utopia that they share. Yet, as I argue, this is not the case for the 'global risk' generations at the beginning of the twenty-first century. These generations are what I call 'generations of side effects'. Their existence and action are not grounded in political action or a new world picture but, first and foremost, in their somewhat pre-embryonic digital existence. The metamorphosis of the world (and the implied change of the frame of reference) has started to change their existence, their understanding of the world, their possibilities of action, and their conception and practice of the social and the political. This change of existence is unfolding without revolt or utopia; it is nothing but the side effect of digitalized modernity turned into the social DNA. These generations incarnate the digital *a priori* – yet not at the end but at the beginning of their socialization. It is not the power of political action that brings out and forms these generations but, metaphorically speaking, their use of mobile phones, which implies different forms and coordinates of communication and living together. The common concept of 'socialization' doesn't capture this any more.

Commonly, socialization means that it is the task of the older generation in the family, school and other institutions to introduce the younger generation into the existing social and political order. As Talcott Parsons (1951) stressed, it is this kind of socialization that ensures that the order of society and politics is stabilized and reproduced over time. An essential prerequisite is that parents and the older generation know and are able to show the young the way, which, in turn, stabilizes their legitimacy and existing hierarchies within the relationship between generations in the family and society.

This model, which allows only for social transformation, breaks down under the conditions of the metamorphosis of the world. This order has been abrogated. Of course, there are still fields in which parents know better. But there are increasingly fields in which this is no longer the case – in fact, in which the roles are reversed: the younger generation turns into the teacher of the older, showing the elderly the way forward. Yet, this happens in a somewhat defensive manner. On the one side, this defensive manner is due to the fact that the younger generation depends socially and materially on the older generation. On the other, it is because this younger generation is unideological and without a clear understanding as to the right way forward; it knows what doesn't work any more without knowing what does work, how it might work and where it leads.

On the world stage of the fight between generations the roles are clearly distributed: the elderly are the *Neanderthals* and the young, global generation are members of *Homo cosmopoliticus*. They are the ones for whom the metamorphosis has become second nature, while the older generations experience it as a threat to their existence. The elderly were born as human beings but, as in Kafka's novel *Metamorphosis* (1915), woke up one morning as insects called the 'digital illiterate'. The young generations, on the contrary, were already born as 'digital beings'. What has been packed into the magic word 'digital' has become part of their 'genetic outfit'.

The generations of the *Homo cosmopoliticus* are still weak and still inferior in the fight between father and daughter and father and son. Their public protests are still dismissed – not least because they are not unified by an idea of a better future for which they could fight against and/or with the older generations. But they become stronger – not least because the *Neanderthals* gradually become extinct. They can only reproduce as digital humans. As we will see further below, the positions of the older generations and those of the *Homo cosmopoliticus* within world risk society are radically different. Already today the global risk generations are better interlinked across borders and more open to the world and its self-destructive potential. In terms of managing everyday life, the *Homo cosmopoliticus* is superior to the *Neanderthals*. Those for whom the metamorphosis of the world has become second nature develop – if all goes well – a competence that enables them to manage life between 'here' and 'there', a life full of avoidance and conciliation, and an ability to deal with contradictions. Yet, the *Neanderthals* revolt. They defend their authority against the *Homo cosmopoliticus*.

As such, the difference between the national outlook and the cosmopolitan outlook turns into a conflict between generations, which is manifest in a clash of generations inside and outside of the family. The case of the migrant family, which lives simultaneously in the West and in other parts of the world, is symptomatic. The daughters can outplay their parents with reference to the law, while the parents literally live in a different world. They epitomize a different understanding of family and a different role of the state. In the West, the law deposes the patriarchal hierarchy within the family. Yet, while a metamorphosis of family order is unfolding, this does not mean that this metamorphosis reaches the thinking of all family members. Families in the West are embedded within a normative system that includes equality of men and women, prohibition of rape within marriage and free choice of partner, all of which are things – in fact, imperatives – that appear strange or even threatening to different family worlds.

Respect, hierarchy and authority turn into the helplessness of the 'divine' rulers and judges of family issues – the man, the father, loses his position, is left behind, and is thrown on the scrap heap. As mentioned earlier, this is not triggered by a revolutionary practice but unfolds behind the façade of continuity in the form of the empowerment of the younger generation and the disempowerment of the older generation: it is a subtle and surreptitious process.

Consequently, memory and the idea of education change too. The internet constitutes something like a memory of all, a collective memory. All libraries of the world, all the information and knowledge that they hold, are accessible to the individual click. In the internet everybody can gain the knowledge they never experienced. There is much criticism of and concern about the fragmented, unorganized, context-free nature of this knowledge – indeed, the danger of drowning in this ocean of (non-)knowledge. Yet, it stands for a metamorphosis that we are still not able to grasp fully. On the one side, the relationship between the teacher and the student is dissolved, even reversed. On the other, the elderly bemoan the supposed downfall of the high value of education and knowledge. But this overlooks the ambivalence of metamorphosis. The concept of education is traditionally oriented inwards.

The philosopher Johann Gottlieb Fichte captured this circular movement of self-knowing consciousness of the world that revolves around itself in the formula 'The ego posits itself'. By this, he (and many other great philosophers) meant that we can use consciousness to traverse and explore consciousness, and in this way to identify the basic characteristics of the world – the transcendental categories of space and time, I and we, society and nature, nation and morality.

This is a grand thought: to find a foothold against the world within oneself. It continues to exert a fascination in the social sciences to this day in works such as those of

Niklas Luhmann and Jürgen Habermas. But it also rests on a magnificent confusion between consciousness and world, between system and world society or world politics. The dire implication is that one need only study oneself in order to understand the world. You don't need to sally forth, you can stay at home and turn in on yourself; you don't have to go out into the world, and you don't have to learn to see the world through the eyes of others in order to understand the world and yourself. This wonderfully frivolous and comfortable error is intrinsic to 'knowledge' that derives its power from the academic claim of self-referential self-reflection (autopoiesis). In this way, culture becomes – and ennobles – narrow-mindedness.

The digital availability of knowing everything oneself (even if this knowledge is not used or if it is turned into its opposite) enforces, or at least enables, a change of horizon. It forces us to move beyond acquirable knowledge – hence, at least incipiently, to see the world through the eyes of others.

2   United in Decline

The metamorphosis of generation and the generation of metamorphosis has to be developed in a sociology of *historical time* that stands beyond the idea of linearity and chronology. At the heart of this thinking is the idea of a coexistence of what could be called 'time-worlds'. This means that the older generations and the younger ones are contemporaries but do not live in the 'same time'. There is no homogeneous similarity. And this, again, is a modus of metamorphosis which has been named by Karl Mannheim and Wilhelm Pinder 'non-temporality of contemporaneous'.

The art historian Pinder (1926) argues against the idea that we can observe relative homogeneous art and style epochs in clear-cut stages. He suggests that, at every moment in time, art-historical epochs and styles exist simultaneously and next to each other. He rejects, so to speak, the idea of 'methodological art epochism' – e.g., the idea that art styles exist and can be studied as closed historical units. With that, he counters in the field of art the idea of evolution and progress according to which one epoch is replaced by another. Pinter's ideas are similar to what later became established as a notion of postmodern eclecticism, at the heart of which are the ideas of deconstruction and liquidation.

When we use the idea of 'non-temporality of contemporaneous' to look at the emergence of global generations, we can, in fact, see some postmodern eclecticism here as well, some dissolution and illusion, but only if we stick to old frames of reference. If we don't stick to these old frames of reference, we see that there are significant variations and fragmentations of global generations, which imply the interplay and confrontation between different horizons and worldviews – as seen, for example, in Occupy, the Arab Spring, generations of young, jobless Southern Europeans, and 'home-grown' fundamentalists.

But this doesn't exclude the fact that there are shared sensibilities and a shared sense of problems and global risks. This, of course, does not imply an equal common response. Even more than that, the sense of the problem differs between different sectors, styles, perceptions, histories and patterns of action.

What this implies is that the understanding of global risk generations can be deduced neither from the biological chronology of generations nor from the idea of a unity of global generations, based, for example, on the shared experience of globality. Being global risk generations certainly does not mean that a worldwide convergence of social situations is taking place. On the contrary: the diversity and inequality of life situations and chances are all too visible, and that is precisely what produces particular tensions and explosive forces.

The normative horizons of global risk generations may be globalized – but they are simultaneously characterized by sharp dividing lines and conflicts. There is above all the economic gap that separates inhabitants of the so-called West from the so-called Rest, the gap in material resources, positions and opportunities of access, which is also evident in the race for the icons of global consumption.

In order to describe the situations and positions of different fractions of younger people in world risk society combining the distribution of goods and bads with the horizons of cultural diversity, we need a new concept for cosmopolitan cross-border research: *generational constellations* (Beck and Beck-Gernsheim 2009).

This is because we have to replace the frame of reference of social change with the frame of reference of metamorphosis. We can no longer deal, as was usual until now, with the single 'generation', understood as existing within the boundaries of the nation-state. 'Generational constellations' represent the cosmopolitan outlook (methodological cosmopolitanism). This implies that the scheme and outlines of unequal positions (such as classes, nations, centre and periphery) are unsuited to representing the inequality of the positions of the risk generations at the beginning of the twenty-first century; new ones are not yet in sight or have not been empirically tested. The following dimensions overlap and interpenetrate in the diagnostic concept 'generational constellations': quantitative, demographic dimension: age polarization; material inequalities: education and labour market position, as well as risk positions and ethnic-cultural diversities.

In order to understand the diverse generational constellations, it is necessary not only to look at the distribution of goods and bads but also to take account of the fact that the principles and expectations of *equality* are spreading worldwide. One important dimension of generational constellations is that there are now normative horizons of equality putting pressure on the existing structures and institutions of global inequality. In this way the nation-state legitimation of transnational or global inequality is beginning to crumble. Even if we have an increase in social inequality globally, inside or outside the nation-state, this will not introduce political conflicts as long as there is not a global expectation of equality. This is because social inequalities do not lead to conflict if the rich become richer and the poor become poorer. It triggers conflict only if established social norms and expectations regarding equality – especially human rights – spread. If one wants to understand the situation of the younger generation, one needs to start by bringing the postcolonial discourse of equality centre stage: in the era of colonial rule the inferiority of others – 'natives', 'savages' – had been defined as (more or less) naturally given. Postcolonial discourse has divested such assumptions of any legitimation. Global risks have similar effects: they intensify worldwide social relations that even in the old 'periphery' are shaping events in the 'old centre', and vice versa. Global risks are therefore no longer processes of uni-directional imperialism. Instead, they are disorderly and chaotic. The global spread of risk, however uneven and sporadic, has produced the global spread of manufactured uncertainties or, to put it another way, the generation of manufactured uncertainties.

The dualism between human rights and national rights of citizens is relativized: now a guarantee of human rights has been normatively prescribed at ever more levels – e.g., in the United Nations Universal Declaration of Human Rights, in the EU treaties and in the constitutions of many nation-states. Such institutionalized norms make it increasingly difficult to distinguish between citizens and non-citizens, nationals and foreigners, and to grant certain rights exclusively to some and not to others. This spreading of norms and expectations of equality has far-reaching consequences for the younger generations. The inequality between the haves and the have-nots, between the rich populations of the world and the remaining world, is no longer accepted as fate but called into question, even if only one-sidedly: it is the others, the excluded, the inhabitants of distant lands and continents, who are beginning to rebel against social inequality – through hopes and dreams of migration, which they are translating into practical activity.

From here we can see that the 'globality' of the different fractions and constellations of the global risk generations is very different: it is definitely not the Western but, rather, the non-Western generational fractions which are rising up against the inequalities across nation-state borders, staking a claim to equality. 'I want in!' is the watchword for these worldwide younger generations standing at the gates of Western societies and vigorously rattling the bars.

A further dimension of this generational constellation refers to the striking mismatch between higher education and unemployment. What we see is that, in many countries of the world, we have the best ever educated generation, which, however, is threatened by a hitherto unknown degree of unemployment. In addition to this we have the risk regime of labour, which spreads globally. For a long time it was believed that a precarious form of employment, which exists in semi-industralized Latin American countries, was something like a pre-modern left-over in the global North, which would gradually diminish and disappear in the transition from an industrial to a service-based society. At the beginning of the twenty-first century we witness the opposite development: precarious 'multi-activity' – previously primarily an indicator of women's work – is a rapidly spreading developmental variant of late employment societies that are running out of attractive, highly skilled and well-paid full-time employment.

This transformation of the world of work affects young people in a particularly severe way. The experience of this generation painfully brings together what used to be exclusive: the best education but the worst prospects in the labour market. At the heart of the global protest movement a new social figure arises: the *no-future graduate* of the generation *precarité*.

Two conclusions may be drawn from these and similar findings. First, the increasing insecurity, which is becoming the basic experience of the younger generation, is not a local, regional or national phenomenon. Rather, this insecurity turns into a key experience of the risk generations, transcending borders, a shared experience which we can sum up in the words *united in decline*.

Beyond that, there is a paradoxical, explosive simultaneity to be discovered here. While in the 'First World', and especially for younger people there, the risks and insecurities of life are growing, the countries that constitute it remain the dream destination for many of the young in the poor regions of the globe. Consequently the existential fears of the former are going to encounter the hopes for the future of the latter. On the one side, a 'generation *less*', which, measured by preceding decades, has to accept material losses; on the other, a 'generation *more*', which, motivated by images of an affluent 'First World', wants to share in that wealth. And both – and this is the crucial point – are fractions of the global generations. What is already becoming visible today will in future perhaps emerge more dramatically: the outlines of a new global redistribution struggle. One side on the defensive, trying to hold onto the remnants of affluence with laws and frontier barriers; the others setting out, charging against these same frontiers with all their strength, driven by the hope of a better life. The result: a conflict-laden interaction – one fraction of the global risk generations against the other.

3   Outlook

At the end of this discussion of the metamorphosis of the world, what becomes obvious is that the problematic of the metamorphosis of inequality is *the* key issue of the future. First, this is because of the institutionalization of the norms of equality, which means that global inequality can no longer be overlooked because the national perspective, which produced the incomparability between national spaces of inequality, no longer works. Existing inequalities are stripped of their legitimacy and hence become (openly or not) a political scandal. Second, it is because inequality increases, also within the national context. Third, the public resources that could compensate for the increasing inequalities are abolished. Fourth, it is because of the distribution of bads, which produces risk-classes, risk-nations and different kinds and degrees of inequalities. There is a synthesis of poverty, vulnerability and the threats implied in climate change and natural disasters. In sum, the *Neanderthal* and the *Homo cosmopoliticus* are living in a world in which inequality has become socially and politically explosive. The problem of inequality arises today in the context of so-called natural disasters that are in effect human-made, set against a horizon in which equality has been promised for all.

References and Bibliography

Adeola, Francis O. (2000) Cross-National Environmental Injustice and Human Rights Issues, *American Behavioral Scientist* 43(4): 686–706.

Agyeman, Julian, and Bob Evans (2004) 'Just Sustainability': The Emerging Discourse of Environmental Justice in Britain?, *Geographical Journal* 170(2): 155–64.

Anderson, Benedict (2006) *Imagined Communities: Reflections on the Origin and Spread of Nationalism*. 2nd edn, London: Verso.

Banister, David (2008) The Sustainable Mobility Paradigm, *Transport Policy* 15(2): 73–80.

Barber, Benjamin (2013) *If Mayors Ruled the World*. New Haven, CT: Yale University Press.

Bauman, Zygmunt (1989) *Modernity and the Holocaust*. Cambridge: Polity.

Beck, Ulrich (1987) The Anthropological Shock: Chernobyl and the Contours of Risk Society, *Berkeley Journal of Sociology: A Critical Review* 32: 153–65.

Beck, Ulrich ([1986] 1992): *Risk Society: Towards a New Modernity*. London: Sage.

Beck, Ulrich (1997) *The Reinvention of Politics: Rethinking Modernity in the Global Social Order*. Cambridge: Polity.

Beck, Ulrich (1998) Misunderstanding Reflexivity: The Controversy on Reflexive Modernization, in Ulrich Beck (ed.), *Democracy without Enemies*. Cambridge: Polity, pp. 84–102.

Beck, Ulrich (1999) *World Risk Society*. Cambridge: Polity.
Beck, Ulrich ([1997] 2000): *What Is Globalization?* Cambridge: Polity.
Beck, Ulrich (2005) *Power in the Global Age: A New Global Political Economy*. Cambridge: Polity.
Beck, Ulrich (2006) *The Cosmopolitan Vision*. Cambridge: Polity.
Beck, Ulrich (2009) *World at Risk*. Cambridge: Polity.
Beck, Ulrich (2010) Remapping Social Inequalities in an Age of Climate Change: For a Cosmopolitan Renewal of Sociology, *Global Networks* 10(2): 165–81.
Beck, Ulrich (2011) Cosmopolitanism as Imagined Communities of Global Risk, in Edward A. Tiryakian (ed.), 'Imagined Communities in the 21st Century', *American Behavioral Scientist* 55(10): 1346–61 [special issue].
Beck, Ulrich (2013a) *German Europe*. Cambridge: Polity.
Beck, Ulrich (2013b) Why 'Class' Is too Soft a Category to Capture the Explosiveness of Social Inequality at the Beginning of the Twenty-First Century, *British Journal of Sociology* 64(1): 63–74.
Beck, Ulrich (2014) How Climate Change Might Save the World: Metamorphosis, *Harvard Design Magazine* 39: 88–98.
Beck, Ulrich (2015) Emancipatory Catastrophism: What Does it Mean to Climate Change and Risk Society?, *Current Sociology* 63(1): 75–88.
Beck, Ulrich, and Elisabeth Beck-Gernsheim (2009) Global Generations and the Trap of Methodological Nationalism: For a Cosmopolitan Turn in the Sociology of Youth and Generation, *European Sociological Review* 25(1): 25–36.
Beck, Ulrich, and Elisabeth Beck-Gernsheim (2014) *Distant Love: Personal Life in the Global Age*. Cambridge: Polity.
Beck, Ulrich, and Daniel Levy (2013) Cosmopolitanized Nations: Re-Imagining Collectivity in World Risk Society, *Theory, Culture & Society* 30(2): 3–31.
Beck, Ulrich, and Peter Wehling (2012) The Politics of Non-Knowing: An Emerging Area of Social and Political Conflict in Reflexive Modernity, in Fernando Domínguez Rubio and Patrick Baert (eds), *The Politics of Knowledge*. London: Routledge, pp. 33–57.
Beck-Gernsheim, Elisabeth (2014) Die schöne neue Welt der Fortpflanzung, in Martina Löw (ed.), *Vielfalt und Zusammenhalt: Verhandlungen des 36. Kongresses der Deutschen Gesellschaft für Soziologie*. Frankfurt: Campus, pp. 161–72.

Beck-Gernsheim, Elisabeth (2015) Danish Sperm and Indian Wombs, in Jean-Daniel Rainhorn and Samira El Boudamoussi (eds), *New Cannibal Markets: Globalization and Commodification of the Human Body*. Paris: Editions de la Maison des sciences de l'homme, pp. 95–103.

Blank, Yishai (2006) The City and the World, *Columbia Journal of Transnational Law* 44(3): 868–931.

Bloch, Ernst ([1954] 1995) *The Principle of Hope*. Cambridge, MA: MIT Press.

Blok, Anders (2012) Greening Cosmopolitan Urbanism? On the Transnational Mobility of Low-Carbon Formats in Northern European and East Asian Cities, *Environment and Planning A* 44(10): 2327–43.

Blok, Anders (2013) Worlding Cities through their Climate Projects? Eco-Housing Assemblages, Cosmopolitics and Comparisons, *CITY* 18(3): 269–86.

Blok, Anders (2015) Towards Cosmopolitan Middle-Range Theorizing: A Metamorphosis in the Practice of Social Theory?, *Current Sociology* 63(1): 110–14.

Blok, Anders, and Robin Tschötschel (2015) World Port Cities as Cosmopolitan Risk Community: Mapping Climate Policy Experiments in Europe and East Asia, under review for *Environment and Planning C*.

Bourdieu, Pierre (1977) *Outline of a Theory of Practice*. Cambridge: Cambridge University Press.

Bourdieu, Pierre (1984) *Distinction: A Social Critique of the Judgement of Taste*. Cambridge, MA: Harvard University Press.

Bourdieu, Pierre (1990) *The Logic of Practice*. Stanford, CA: Stanford University Press.

Brenner, Neil (ed.) (2014) *Implosions/Explosions: Towards a Study of Planetary Urbanization*. Berlin: Jovis.

Broto, Vanesa C., and Harriet Bulkeley (2013) A Survey of Urban Climate Change Experiments in 100 Cities, *Global Environmental Change* 23(1): 92–102.

Bulkeley, Harriet (2013) *Cities and Climate Change*. London: Routledge.

Bulkeley, Harriet, Liliana Andonova, Karin Bäckstrand, Michele Betsill, Daniel Compagnon, Rosaleen Duffy, Ans Kolk, Matthew Hoffmann, David Levy, Peter Newell, Tori Milledge, Matthew Paterson, Philipp Pattberg and Stacy Vandeveer (2012) Governing

Climate Change Transnationally: Assessing the Evidence from a Database of Sixty Initiatives, *Environment and Planning C* 30(4): 591–612.

Bullard, Robert Doyle, and Beverly Wright (2009) *Race, Place, and Environmental Justice after Hurricane Katrina: Struggles to Reclaim, Rebuild, and Revitalize New Orleans and the Gulf Coast*. Boulder, CO: Westview Press.

Campbell, Denis, and Nicola Davison (2012) Illegal Kidney Trade Booms as New Organ is 'Sold Every Hour', *The Guardian*, 27 May, www.theguardian.com/world/2012/may/27/kidney-trade -illegal-operations-who.

Castells, Manuel (1996) *The Rise of the Network Society*. Oxford: Blackwell.

Chaix, Basile, Susanna Gustafsson, Michael Jerrett, Håkan Kristersson, Thor Lithman, Åke Boalt and Juan Merlo (2006) Children's Exposure to Nitrogen Dioxide in Sweden: Investigating Environmental Justice in an Egalitarian Country, *Journal of Epidemiology and Community Health* 60(3): 234–41.

Chiapello, Eve (2013) Capitalism and its Criticisms, in Paul du Gay and Glenn Morgan (eds), *New Spirits of Capitalism? Crises, Justifications and Dynamics*. Oxford: Oxford University Press, pp. 60–81.

Crutzen, Paul J. (2006) The 'Anthropocene', in Eckart Ehlers and Thomas Krafft (eds), *Earth System Science in the Anthropocene*. New York: Springer, pp. 13–18.

Cutter, Susan L., and Christopher T. Emrich (2006) Moral Hazard, Social Catastrophe: The Changing Face of Vulnerability along the Hurricane Coasts, *Annals of the American Academy of Political and Social Science* 604(1): 102–12.

Cutter, Susan L., Bryan J. Boruff, and W. Lynn Shirley (2003) Social Vulnerability to Environmental Hazards, *Social Science Quarterly* 84(1): 242–61.

Dewey, John ([1927] 1954) *The Public and its Problems*. Chicago: Swallow Press.

Dixon, Jacqueline, and Maano Ramutsindela (2006) Urban Resettlement and Environmental Justice in Cape Town, *Cities* 23(2): 129–39.

Eisenstadt, Shmuel N. (1986) Introduction: The Axial Age Breakthroughs – their Characteristics and Origins, in Shmuel N. Eisenstadt (ed.), *The Origins and Diversity of Axial Age Civilisations*. Albany: State University of New York Press, pp. 1–25.

Fan, Mei-Fang (2006) Environmental Justice and Nuclear Waste Conflicts in Taiwan, *Environmental Politics* 15(3): 417–34.

Fichtner, Ullrich (2014) The Grapes of Wrath: France's Great Wines are Feeling the Heat, *Spiegel Online International*, 30 October, www.spiegel.de/international/zeitgeist/climate-change-threatens-french-viticulture-a-1000113.html.

Fielding, Jane, and Kate Burningham (2005) Environmental Inequality and Flood Hazard, *Local Environment* 10(4): 1–17.

Fischer, David (1997) *History of the International Atomic Energy Agency: The First Forty Years*. Vienna: IAEA.

Foucault, Michel (1980) *Power/Knowledge: Selected Interviews and Other Writings, 1972–1977*. Brighton: Harvester Press.

Grusin, Richard (2010) *Premediation: Affect and Mediality after 9/11*. Basingstoke: Palgrave Macmillan.

Guyer, Jane I. (2007) Prophecy and the Near Future: Thoughts on Macroeconomic, Evangelical and Punctuated Time, *American Ethnologist* 34(3): 409–21.

Habermas, Jürgen (1987) *The Theory of Communicative Action*. Cambridge: Polity.

Habermas, Jürgen (1997) Kant's Idea of Perpetual Peace, with the Benefit of Two Hundred Years' Hindsight, in James Bohman and Matthias Lutz-Bachmann (eds), *Perpetual Peace: Essays on Kant's Cosmopolitan Ideal*. Cambrdige, MA: MIT Press, pp. 113–53.

Hamilton, Clive (2010) *Requiem for a Species*. London: Earthscan.

Harich, Wolfang (1975) *Kommunismus ohne Wachstum? Babeuf und der 'Club of Rome' [Communism without Growth: Babeuf and the Club of Rome]*. Reinbek bei Hamburg: Rowohlt.

Hillman, Michael J. (2006) Situated Justice in Environmental Decision-Making: Lessons from River Management in Southeastern Australia, *Geoforum* 37(5): 695–707.

Hobbs, Dick (2013) *Lush Life: Constructing Organized Crime in the UK*. Oxford: Oxford University Press.

Hodson, Mike, and Simon Marvin (2010) *World Cities and Climate Change: Producing Urban Ecological Security*. Maidenhead: Open University Press.

Hondagneu-Sotelo, Pierrette, and Ernestine Avila (1997) 'I am Here, but I am There': The Meanings of Latina Transnational Motherhood, *Gender & Society* 11(5): 548–71.

Inhorn, Marcia C. (2003) *Local Babies, Global Science: Gender, Religion and in vitro Fertilization in Egypt*. London: Routledge.

Jaspers, Karl (1953) *The Origin and Goal of History*. New Haven, CT: Yale University Press.

Joas, Hans (1996) *The Creativity of Action*. Chicago: University of Chicago Press.

Joss, Simon, Robert Cowley and Daniel Tomozeiu (2013) Towards the 'Ubiquitous Eco-City': An Analysis of the Internationalization of Eco-City Policy and Practice, *Urban Research & Practice* 6(1): 54–74.

Kafka, Franz ([1915] 2014) *The Metamorphosis*. New York: W. W. Norton.

Kant, Immanuel ([1795] 1972) *Perpetual Peace: A Philosophical Essay*. New York: Garland.

Köhler, Benedikt (2006) *Soziologie des Neuen Kosmopolitismus*. Wiesbaden: VS Verlag für Sozialwissenschaften.

Krüger, Michael (2009) Menschenrechte und Marillenknödel, *Süddeutsche Zeitung*, 18 June, p. 13.

Kuchinskaya, Olga (2014) *The Politics of Invisibility: Public Knowledge about Radiation Health Effects after Chernobyl*. Cambridge, MA: MIT Press.

Kuhn, Thomas S. (1962) *The Structure of Scientific Revolutions*. Chicago: University of Chicago Press.

Kurasawa, Fuyuki (2007) *The Work of Global Justice: Human Rights as Practices*. Cambridge: Cambridge University Press.

Kurasawa, Fuyuki (2014) In Praise of Ambiguity: On the Visual Economy of Distant Suffering, in Ratiba Hadj-Moussa and Michael Nijhawan (eds), *Suffering, Art, and Aesthetics*. Basingstoke: Palgrave Macmillan, pp. 23–50.

Lakatos, Imre (1978) *The Methodology of Scientific Research Programmes*, ed. John Worrall and Gregory Currie. Cambridge: Cambridge University Press.

Latour, Bruno (2011) Waiting for Gaia: Composing the Common World through Arts and Politics, lecture delivered at the French Institute in London, November, http://www.bruno-latour.fr/sites/default/files/124-GAIA-LONDON-SPEAP_0.pdf.

Lee, Taedong, and Susan van de Meene (2012) Who Teaches and Who Learns? Policy Learning through the C40 Cities Climate Network, *Policy Sciences* 45(3): 199–220.

Lin, Xiaochun, and Jun Yang (2009) In Dealing with Climate Change, China Shows its Attitudes, *People's Daily*, 18 December, p. 2.

London, Leslie (2003) Human Rights, Environmental Justice, and the Health of Farm Workers in South Africa, *International Journal of Occupational and Environmental Health* 9(1): 59–68.

Luhmann, Niklas (1995) *Social Systems*. Stanford, CA: Stanford University Press.

Luhmann, Niklas (2012) *Theory of Society*, Volume 1. Stanford, CA: Stanford University Press.

Maantay, Juliana, and Andrew Maroko (2009) Mapping Urban Risk: Flood Hazards, Race, & Environmental Justice in New York, *Applied Geography* 29(1): 111–24.

Mannheim, Karl ([1928] 1952) The Problem of Generations, in Karl Mannheim, *Essays on the Sociology of Knowledge*, ed. Paul Kecskemeti. London: Routledge & Kegan Paul, pp. 276–322.

Mao, Zhifei (2014a) Cosmopolitanism and the Media Construction of Risk, unpublished working paper, Hong Kong.

Mao, Zhifei (2014b) Cosmopolitanism and Global Risk: News Framing of the Asian Financial Crisis and the European Debt Crisis, *International Journal of Communication* 8: 1029–48.

Merton, Robert K. (1968) *Social Theory and Social Structures*. New York: Free Press.

Mills, C. Wright (1959) *The Sociological Imagination*. New York: Oxford University Press.

Moore, Henrietta L., and Sabine Selchow (2012) Global Civil Society and the Internet 2012: Time to Update our Perspective, in Mary Kaldor, Henrietta L. Moore and Sabine Selchow (eds), *Global Civil Society 2012: Ten Years of Critical Reflection*. Basingstoke: Palgrave Macmillan, pp. 28–40.

Mythen, Gabe (2005) From 'Goods' to 'Bads'? Revisiting the Political Economy of Risk, *Sociological Research Online* 10(3), www.socresonline.org.uk/10/3/mythen.html.

Mythen, Gabe (2014) *Understanding the Risk Society: Crime, Security and Justice*. Basingstoke: Palgrave Macmillan.

Newell, Peter (2005) Race, Class and the Global Politics of Environmental Inequality, *Global Environmental Politics* 5(3): 70–94.

Nietzsche, Friedrich W. (1920) *The Antichrist*. New York: Knopf.

Oliver-Smith, Anthony (1996) Anthropological Research on Hazards and Disasters, *Annual Review of Anthropology* 25: 303–28.

Omer, Itzhak, and Udi Or (2005) Distributive Environmental Justice in the City: Differential Access in Two Mixed Israeli Cities, *Tijdschrift voor Economische en Sociale Geografie* 96(4): 433–43.

Pariser, Eli (2011) *The Filter Bubble: How the New Personalized Web is Changing What We Read and How We Think*. New York: Penguin Press.

Parsons, Talcott (1951) *The Social System*. Glencoe, IL: Free Press.

Pascal, Blaise ([1670] 1958) *Pascal's Pensées*. New York: E. P. Dutton; www.gutenberg.org/ebooks/18269?msg=welcome_stranger #SECTION_III.

Pearce, Jamie, Simon Kingham and Peyman Zawar-Reza (2006) Every Breath You Take? Environmental Justice and Air Pollution in Christchurch, New Zealand, *Environment and Planning A* 38(5): 919–38.

Pellow, David, Tamara Steger and Rebecca McLain (2005) *Proceedings from the Transatlantic Initiative to Promote Environmental Justice Workshop*. Central European University, Budapest, Hungary, 27–30 October, http://archive.ceu.hu/pub lications/pellow/20015/42818.

Petryna, Adriana (2003) *Life Exposed: Biological Citizens after Chernobyl*. Princeton, NJ: Princeton University Press.

Phillips, Brenda D., Deborah S. K. Thomas, Alice Fothergill, and Lynn Blinn-Pike (eds) (2010) *Social Vulnerability to Disasters*. Boca Raton, FL: CRC Press.

Piketty, Thomas (2014) *Capital in the Twenty-First Century*. Cambridge, MA: Belknap Press of Harvard University Press.

Pinder, Wilhelm (1926) *Das Problem der Generation in der Kunstgeschichte Europas*. Berlin: Frankfurter Verlags-Anstalt.

Russill, Chris, and Zoe Nyssa (2009) The Tipping Point Trend in Climate Change Communication, *Global Environmental Change* 19(3): 336–44.

Sassen, Saskia (2009) Cities Are at the Center of our Environmental Future, *Sapiens* 2(3): 1–8, http://sapiens.revues.org/948.

Sassen, Saskia (2014) Recovering the City Level in the Global Environmental Struggle, in Stewart Lockie, David A. Sonnenfeld and Dana R. Fisher (eds), *Routledge International Hand-*

book of Social and Environmental Change. London: Routledge, pp. 170–8.

Selchow, Sabine (2014) Security Policy and (Global) Risk(s), in Mary Kaldor and Iavor Rangelov (eds), *The Handbook of Global Security Policy*. Chichester: Wiley-Blackwell, pp. 68–84.

Sheller, Mimi, and John Urry (2000) The City and the Car, *International Journal of Urban and Regional Research* 24(4): 737–57.

Slater, Don (2013) *New Media, Development and Globalization: Making Connections in the Global South*. Cambridge: Polity.

Steger, Tamara (ed.) (2007) *Making the Case for Environmental Justice in Central & Eastern Europe*. Budapest: CEU Center for Environmental Policy and Law (CEPL), Health and Environment Alliance (HEAL), and Coalition for Environmental Justice, www.env-health.org/IMG/pdf/28-_Making_the_case _for_environmental_justice_in_Europe.pdf.

Steiner, Benjamin (2015) *Nebenfolgen in der Geschichte: Eine historische Soziologie reflexiver Modernisierung*. Berlin: De Gruyter.

Stewart, Quincy Thomas, and Rashawn Ray (2007) Hurricane Katrina and the Race Flood, *Race, Gender and Class* 14(1/2): 38–59.

Swyngedouw, Erik (2010) Apocalypse Forever? Post-Political Populism and the Spectre of Climate Change, *Theory, Culture & Society* 27(2/3): 213–32.

Szerszynski, Bronislaw (2010) Reading and Writing the Weather: Climate Technics and the Moment of Responsibility, *Theory, Culture & Society* 27(2/3): 9–30.

Thorsen, Line Marie (2014) Art and Climate Change: Cosmopolitization of Aesthetics /Aesthetics of Cosmopolitization, unpublished manuscript, Copenhagen.

Tollefson, Jeff (2012) Hurricane Sweeps US into Climate-Adaptation Debate, *Nature* 491: 167–8.

Ueland, Jeff, and Barney Warf (2006) Radicalized Topographies: Altitude and Race in Southern Cities, *Geographical Review* 96(1): 50–78.

UNDP (United Nations Development Programme) and Committee on the Problems of the Consequences of the Catastrophe at the Chernobyl NPP (2004) *Report: An Information Needs Assessment of the Chernobyl-Affected Population in the*

Republic of Belarus. Minsk: Unipack; http://un.by/pdf/CHE
_OON_ENG.pdf.

Vara, Ana María (2015) A South American Approach to Metamorphosis as a Horizon of Equality: Focusing on Controversies
over Lithium, *Current Sociology* 63(1): 100–4.

Volkmer, Ingrid (2014) *The Global Public Sphere: Public Communication in the Age of Reflective Independence*. Cambridge:
Polity.

Waldman, Ellen (2006) Cultural Priorities Revealed: The Development and Regulation of Assisted Reproduction in the United
States and Israel, *Health Matrix: Journal of Law-Medicine* 16:
65–106.

Walker, Gordon (2009a) Beyond Distributional Proximity: Exploring the Multiple Spatialities of Environmental Justice, *Antipode*
41(4): 614–63.

Walker, Gordon (2009b) Globalizing Environmental Justice: The
Geography and Politics of Frame Contextualization and Evolution, *Global Social Policy* 9(3): 355–82.

Walker, Gordon, and Kate Burningham (2011) Flood Risk, Vulnerability and Environmental Justice: Evidence and Evaluation
of Inequality in a UK Context, *Critical Social Policy* 31(2):
216–40.

Walker, Gordon, Kate Burningham, Jane Fielding, Graham
Smith, Diana Thrush and Helen Fay (2006) *Addressing Envi-
ronmental Inequalities: Flood Risk*. Bristol: Environment
Agency.

Walker, Gordon, John Fairburn, Graham Smith and Gordon
Mitchell (2003) *Environmental Quality and Social Deprivation, Phase II: National Analysis of Flood Hazard, IPC Industries and Air Quality*. Bristol: Environment Agency.

Weber, Max (1922) *Gesammelte Aufsätze zur Wissenschaftslehre
(Collected Essays on Epistemology)*. Tübingen: Mohr.

Wehling, Peter (2006) *Im Schatten des Wissens? Perspektiven der
Soziologie des Nichtwissens*. Konstanz: UVK.

Werrity, Alan, Donald Houston, Tom Ball, Amy Tavendale and
Andrew Black (2007) *Exploring the Social Impacts of Flood
Risk and Flooding in Scotland*. Edinburgh: Scottish Executive
Social Research.

Wilby, R. L. (2007) A Review of Climate Change Impacts on the
Built Environment, *Built Environment* 33(1): 31–45.

Wimmer, Andreas, and Nina Glick Schiller (2002) Methodological Nationalism and Beyond: Nation-State Building, Migration and the Social Sciences, *Global Networks* 2(4): 301–34.

Wimmer, Andreas, and Nina Glick Schiller (2003) Methodological Nationalism, the Social Sciences, and the Study of Migration: An Essay in Historical Epistemology, *International Migration Review* 37(3): 576–610.

Xie, Lianhui (1991) Li Xue's Discussion on the Academic Conference of Climate Change and the Environmental Problem, *People's Daily*, 16 January, p. 3.

Xinhua News Agency (1962) Annual Conference of Chinese Meteorologists, *People's Daily*, 13 August, p. 2.

Xinhua News Agency (1980) Does the Slowing Speed of Earth's Rotation Correlate with Climate Change?, *People's Daily*, 22 March, p. 6.

Xinhua News Agency (1994) United Nations Framework Convention on Climate Change Has Ended: The Developed Countries Must Take More Responsibility, *People's Daily*, 20 February, p. 7.

Yates, Joshua J. (2009) Mapping the Good World: The New Cosmopolitans and our Changing World Picture, *Hedgehog Review* 11(3): 7–27.

Zhang, Jiacheng, and Mingdao Zhu (1973) A Discussion of the Climate Change in Recent Years, *People's Daily*, 21 July, p. 3.

Zhao, Cheng, Fan Tian and Dong Wei (2009) Premier Wen Jiabao attended the 2009 United Nations Climate Change Conference, *People's Daily*, 25 December, p. 1.

Zheng, Sizhong (1979) Two Different Points of View in Regard to Global Climate Change, *People's Daily*, 21 August, p. 6.

Zou, Ji (2007) The Average Emissions of Greenhouse Gas in China are Much Lower than Those in the Developed Countries, *People's Daily*, 29 March, p. 16.

Index

Milton Keynes UK
Ingram Content Group UK Ltd.
UKHW021829160124
436144UK00017B/813